AF595462

Urban Agriculture, Forestry and Green-Blue Infrastructure as "Re-discovered Commons": Bridging Urban-Rural Interface

Urban Agriculture, Forestry and Green-Blue Infrastructure as "Re-discovered Commons": Bridging Urban-Rural Interface

Editors

Ryo Kohsaka
Yuta Uchiyama

MDPI • Basel • Beijing • Wuhan • Barcelona • Belgrade • Manchester • Tokyo • Cluj • Tianjin

Editors

Ryo Kohsaka
Graduate School of
Environmental Studies
Nagoya University
Nagoya
Japan

Yuta Uchiyama
Graduate School of
Environmental Studies
Nagoya University
Nagoya
Japan

Editorial Office
MDPI
St. Alban-Anlage 66
4052 Basel, Switzerland

This is a reprint of articles from the Special Issue published online in the open access journal *Sustainability* (ISSN 2071-1050) (available at: www.mdpi.com/journal/sustainability/special_issues/urban_rural).

For citation purposes, cite each article independently as indicated on the article page online and as indicated below:

LastName, A.A.; LastName, B.B.; LastName, C.C. Article Title. *Journal Name* **Year**, *Volume Number*, Page Range.

ISBN 978-3-0365-1810-7 (Hbk)
ISBN 978-3-0365-1809-1 (PDF)

Contents

MDPI

Editorial

Special Issue: "Urban Agriculture, Forestry and Green-Blue Infrastructure as "Re-Discovered Commons": Bridging Urban-Rural Interface"

Ryo Kohsaka * and Yuta Uchiyama

Graduate School of Environmental Studies, Nagoya University, Furo-cho Chikusa-ku D2-1, Aichi 464-8601, Japan; uchiyama.yuta@k.mbox.nagoya-u.ac.jp
* Correspondence: kohsaka@hotmail.com

Citation: Kohsaka, R.; Uchiyama, Y. Special Issue: "Urban Agriculture, Forestry and Green-Blue Infrastructure as "Re-Discovered Commons": Bridging Urban-Rural Interface". *Sustainability* **2021**, *13*, 5872. https://doi.org/10.3390/su13115872

Received: 16 May 2021
Accepted: 18 May 2021
Published: 24 May 2021

This Special Issue re-explores research topics related to the relationships between urban and rural areas during the COVID-19 pandemic period in 2020 and beyond. We revisit the roles and values of the components of agricultural lands and forestlands—as well as urban ones—from the perspective of green–blue infrastructure. In doing so, we propose that the roles are redefined to reflect the transformations of lifestyles and to underpin values in this era, often referred to as the "new normal." During the pandemic period, several national governments implemented strict lockdown policies to avoid the spread of the virus, including certain democratic countries in Europe. Due to the physical and mental restrictions caused by such policies and risk of infection, citizens have become more aware of the meanings and values of green–blue infrastructure, which comprise different types of green areas and river–lake–coastal networks. As a related international process, in the current ongoing discussion of the post-2020 global biodiversity framework in the Subsidiary Body on Scientific, Technical and Technological Advice (SBSTTA), "access to green areas" is included as one of the targets. These components of green–blue infrastructure play essential roles in urban–rural interfaces and are changing urban–rural relationships in terms of citizens' awareness, activities, and work style. As an example, the portmanteau *workcation*—created from "work" and "vacation"—means doing work in areas traditionally used for vacation; the Japanese government is promoting this work style to curb the spread of the virus in high-density urban work environments.

This pandemic period provides us with unique opportunities to explore the changing components and possible transformations of the urban–rural interface in a relatively short period. The purpose of this Special Issue is to provide scientific evidence for the rediscovery and description of "new normal" urban–rural relationships. In the global urbanization process, rural populations have moved to urban areas; however, the status of urbanization has become more complex, as shrinking cities with depopulation and aging are increasing [1–3]. In this context, new urban–rural interfaces need to be developed, considering the new normal lifestyle. Because of the timing of the pandemic period, not all studies in this Special Issue consider the impact of COVID-19 directly; however, there are crucial topics that need to be considered and reexplored to achieve the goals of this Special Issue.

The following part of this editorial summarizes the findings and discussions of the papers included in the Special Issue that are provided.

A component of the urban–rural interface, green areas, is analyzed in terms of citizens' access, or visitation. In the United States, parks play an essential role in the pandemic period. According to Heo et al. [4], who analyzed citizens' trip data—collated by social media use—park use during this period is an essential activity for citizens. The same study analyzed the relationship between the status of vegetation in parks and individuals' park visitation. However, there was no clear correlation between the two variables. Their findings suggest that further analysis is needed to determine the detailed relationships between the status of vegetation and individuals visiting green areas. The status of vegetation might be related to people's preference of which green areas to visit. Future research is needed to verify the possible influence of vegetation status on citizens' preference for visiting green areas. There have been certain empirical explorations of the public visiting green areas, including those examining the "extinction of experiences" [5–7] over a relatively long period, based on official statistics; however, these types of changes during the COVID-19 pandemic offer new contributions to discuss.

Because of the restrictions on citizens' activities—such as going to parks—under lockdown policies, the status of green area visitation is changing. According to a study conducted by the authors [8], the frequency of visiting green areas tends to be higher for certain demographic characteristics; these include household income, gender, and age. The frequency of citizens with higher household income visiting green areas, for example, tended to be higher, and the frequency of females visiting green areas tended to be higher than that of males. The average age of respondents who frequently visited green areas was lower. The initial survey results are provided in this paper and can serve as a basis for green area management during the pandemic period. Further analysis is needed to identify the holistic characteristics of citizens and their typologies in terms of visiting green areas to determine policy targets or segments. Moreover, a comparative analysis between past trends of visiting green areas [9] and current trends is another topic for future research.

The influence of the COVID-19 pandemic on citizens' behaviors and the status of the environment can be considered using the Driver–Pressure–State–Impact–Response (DPSIR) framework. It has previously been utilized for the assessment of urban biodiversity and other local environments [10–13]. A city biodiversity index was proposed by the government of Singapore, and this index has been applied to cities in different regions of the world [14–17]. Such an index can possibly be a tool for evaluating the impact of the pandemic on biodiversity and its governance.

In the future, methodological considerations also need to be considered. Uchiyama and Kohsaka [18] evaluated the natural resources of residents using GIS-based approaches. These methods can be applied and dimensions can be added to the analysis of residents' attributes regarding their visitation to green areas. Other potential application fields are verbal communication, based on face-to-face discussions. Qualitative data, such as verbal communication data, can be utilized to complement quantitative survey results. Text mining analysis can also be used to examine the integrated analysis of qualitative and quantitative data [19,20].

As a case study to analyze the functions of green areas, Yoo et al. [21] conducted surveys and analyses on their function of reducing particulate matter. Their study identified certain positive impacts of green areas. The results imply the importance of managing the functions of green areas located between urban and rural areas, in addition to green areas' air purification functions in urban areas. In future, a comparative analysis of their functions under different conditions is necessary to obtain robust results to verify the functions of green areas.

Regarding forest management-related issues, proper management is necessary to maintain and enhance the environmental quality of the urban–rural interface. Sustainable production of forest products is one of the main elements of forest management, and scientific evidence is necessary to develop future policies for forest use and to make forest management plans with local stakeholders. Understanding local perceptions and value systems [22–24] is fundamental for participatory forest management. The interaction of all actors in forest management and their knowledge exchange process [25,26] need to be explored to enhance the quality of management in terms of urban–rural interlinkages. In 2019, the Japanese government introduced a national-level forest environment transfer tax. Such schemes related to payment for ecosystem services can possibly be an appropriate measure to facilitate urban–rural collaboration based on evidence that shows the effective use of the tax [27].

The well-being of citizens who are involved in forest management needs to be suitably evaluated, in addition to the evaluation of forest functions. Although environmental factors need to be maintained as essential factors of the urban–rural interface, the well-being of actors who contribute to environmental management must also be considered during the policy-making process. As a possible indicator of subjective well-being, Takahashi et al. [28] proposed relevant indicators by providing the results of a case study to demonstrate the applicability of the indicators. Indicator-based management of forests and evidence-based policy making are necessary to enhance the well-being of both urban and rural citizens.

Regarding the socio-ecological landscape of the urban–rural interface, materials used in buildings are elements that also need to be considered as components of the urban–rural ecosystem. Park et al. [29] point out that the traditional materials used to build houses can accommodate diverse bird species to nest, and that such residential areas can contribute to regional biodiversity management. In the gradation of urban–rural interfaces, residential areas can play an important role in terms of biodiversity and ecosystem conservation, as well as other land use categories, such as agricultural lands and forestlands.

As a component of green–blue infrastructure and river–lake–coastal networks, the blue carbon ecosystem is gaining salience in the international arena. Coastal management requires appropriate collaboration during watershed moments, including urban and rural areas. Furthermore, the blue carbon ecosystem can contribute to a global mitigation approach for climate change, based on carbon stocks. Reducing risk and providing resources for local livelihoods are also included in the functions of the blue carbon ecosystem. Quevedo et al. [30] conducted a comparative analysis of local perceptions in the coral triangle area, which has one of the largest and richest blue carbon ecosystems. Local perceptions can be a baseline for the management of the ecosystem, as their perceptions need to be considered in management as the main actors of management. As the enhancement of their awareness of the ecosystem is a key factor, their perceptions need to be analyzed to obtain scientific evidence to elaborate the schemes for their awareness enhancement [31–33]. The local perceptions of urban and rural people differ [34]; moreover, an analysis of existing policy is also needed to enhance collaboration among local actors and policy interventions [35].

The topics of the papers published in this Special Issue—green areas, forest-related ecosystems and traditions, forest management, and blue carbon—were introduced and discussed. Diverse key topics are discussed in this Special Issue, and an "evidence-based approach" to the urban–rural interface is derived as the common key topic, which can be further discussed in future issues.

Published papers in the Special Issue are:

1. Relationships between Local Green Space and Human Mobility Patterns during COVID-19 for Maryland and California, USA by Seulkee Heo, Chris C. Lim and Michelle L. Bell [4].
2. Access and Use of Green Areas during the COVID-19 Pandemic: Green Infrastructure Management in the "New Normal" by Yuta Uchiyama and Ryo Kohsaka [8].
3. Importance of Urban Green at Reduction of Particulate Matters in Sihwa Industrial Complex, Korea by Sin-Yee Yoo, Taehee Kim, Suhan Ham, Sumin Choi and Chan-Ryul Park [21].
4. Reconstruction of Resin Collection History of Pine Forests in Korea from Tree-Ring Dating by En-Bi Choi, Yo-Jung Kim, Jun-Hui Park, Chan-Ryul Park and Jeong-Wook Seo.
5. Subjective Well-Being as a Potential Policy Indicator in the Context of Urbanization and Forest Restoration by Takuya Takahashi, Yukiko Uchida, Hiroyuki Ishibashi and Noboru Okuda [28].
6. The Functional Traits of Breeding Bird Communities at Traditional Folk Villages in Korea by Chan Ryul Park, Sohyeon Suk and Sumin Choi [29].
7. How Blue Carbon Ecosystems Are Perceived by Local Communities in the Coral Triangle: Comparative and Empirical Examinations in the Philippines and Indonesia by Jay Mar D. Quevedo, Yuta Uchiyama, Kevin Muhamad Lukman and Ryo Kohsaka [30].

Funding: This work was supported by the JSPS KAKENHI Grant Numbers JP16KK0053; JP17K02105; 17K13305; 20K12398; Japan Science and Technology Agency (JST) and Japan International Cooperation Agency (JICA) through the Science and Technology Research Partnership for Sustainable Development Program (SATREPS)—Comprehensive Assessment and Conservation of Blue Carbon Ecosystems and Their Services in the Coral Triangle (Blue CARES) project; JST RISTEX Grant Number JPMJRX20B3; Asia-Pacific Network for Global Change Research Grant Number CBA2020-05SY-Kohsaka; Kurita Water and Environment Foundation [20C002]; and Foundation for Environmental Conservation Measures, Keidanren (2020).

Conflicts of Interest: The authors declare no conflict of interest.

References

1. Döringer, S.; Uchiyama, Y.; Penker, M.; Kohsaka, R. A meta-analysis of shrinking cities in Europe and Japan. Towards an integrative research agenda. *Eur. Plan. Stud.* **2020**, *28*, 1693–1712. [CrossRef]
2. Hollander, J.B.; Pallagst, K.; Schwarz, T.; Popper, F.J. Planning shrinking cities. *Prog. Plan.* **2009**, *72*, 223–232.
3. Martinez-Fernandez, C.; Audirac, I.; Fol, S.; Cunningham-Sabot, E. Shrinking Cities: Urban Challenges of Globalization. *Int. J. Urban Reg. Res.* **2012**, *36*, 213–225. [CrossRef] [PubMed]
4. Heo, S.; Lim, C.C.; Bell, M.L. Relationships between Local Green Space and Human Mobility Patterns during COVID-19 for Maryland and California, USA. *Sustainability* **2020**, *12*, 9401. [CrossRef]
5. Imai, H.; Nakashizuka, T.; Kohsaka, R. An analysis of 15 years of trends in children's connection with nature and its relationship with residential environment. *Ecosyst. Health Sustain.* **2018**, *4*, 177–187. [CrossRef]
6. Miller, J.R. Biodiversity conservation and the extinction of experience. *Trends Ecol. Evol.* **2005**, *20*, 430–434. [CrossRef]
7. Soga, M.; Gaston, K.J. Extinction of experience: The loss of human-nature interactions. *Front. Ecol. Environ.* **2016**, *14*, 94–101. [CrossRef]
8. Uchiyama, Y.; Kohsaka, R. Access and Use of Green Areas during the COVID-19 Pandemic: Green Infrastructure Management in the "New Normal&rdquo. *Sustainability* **2020**, *12*, 9842. [CrossRef]
9. Aikoh, T.; Abe, R.; Kohsaka, R.; Iwata, M.; Shoji, Y. Factors Influencing Visitors to Suburban Open Space Areas near a Northern Japanese City. *Forests* **2012**, *3*, 155–165. [CrossRef]
10. Kelble, C.R.; Loomis, D.K.; Lovelace, S.; Nuttle, W.K.; Ortner, P.B.; Fletcher, P.; Cook, G.S.; Lorenz, J.J.; Boyer, J.N. The EBM-DPSER Conceptual Model: Integrating Ecosystem Services into the DPSIR Framework. *PLoS ONE* **2013**, *8*, e70766. [CrossRef]
11. Kohsaka, R. Developing biodiversity indicators for cities: Applying the DPSIR model to Nagoya and integrating social and ecological aspects. *Ecol. Res.* **2010**, *25*, 925–936. [CrossRef]
12. Quevedo, J.M.D.; Uchiyama, Y.; Kohsaka, R. A blue carbon ecosystems qualitative assessment applying the DPSIR framework: Local perspective of global benefits and contributions. *Mar. Policy* **2021**, *128*, 104462. [CrossRef]
13. Sun, S.; Wang, Y.; Liu, J.; Cai, H.; Wu, P.; Geng, Q.; Xu, L. Sustainability assessment of regional water resources under the DPSIR framework. *J. Hydrol.* **2016**, *532*, 140–148. [CrossRef]
14. Deslauriers, M.R.; Asgary, A.; Nazarnia, N.; Jaeger, J.A. Implementing the connectivity of natural areas in cities as an indicator in the City Biodiversity Index (CBI). *Ecol. Indic.* **2018**, *94*, 99–113. [CrossRef]
15. Kohsaka, R.; Pereira, H.M.; Elmqvist, T.; Chan, L.; Moreno-Peñaranda, R.; Morimoto, Y.; Inoue, T.; Iwata, M.; Nishi, M.; Mathias, M.D.L.; et al. Indicators for Management of Urban Biodiversity and Ecosystem Services: City Biodiversity Index. In *Urbanization, Biodiversity and Ecosystem Services: Challenges and Opportunities*; Springer: Dordrecht, The Netherlands, 2013; pp. 699–718.
16. Kohsaka, R.; Uchiyama, Y. Motivation, Strategy and Challenges of Conserving Urban Biodiversity in Local Contexts: Cases of 12 Municipalities in Ishikawa, Japan. *Procedia Eng.* **2017**, *198*, 212–218. [CrossRef]
17. Uchiyama, Y.; Kohsaka, R. Application of the City Biodiversity Index to populated cities in Japan: Influence of the social and ecological characteristics on indicator-based management. *Ecol. Indic.* **2019**, *106*, 105420. [CrossRef]
18. Uchiyama, Y.; Kohsaka, R. Cognitive value of tourism resources and their relationship with accessibility: A case of Noto region, Japan. *Tour. Manag. Perspect.* **2016**, *19*, 61–68. [CrossRef]
19. Kohsaka, R.; Matsuoka, H. Analysis of Japanese municipalities with Geopark, MAB, and GIAHS certification: Quantitative approach to official records with text-mining methods. *SAGE Open* **2015**, *5*. [CrossRef]
20. Wetts, R. In climate news, statements from large businesses and opponents of climate action receive heightened visibility. *Proc. Natl. Acad. Sci. USA* **2020**, *117*, 19054–19060. [CrossRef] [PubMed]
21. Yoo, S.-Y.; Kim, T.; Ham, S.; Choi, S.; Park, C.-R. Importance of Urban Green at Reduction of Particulate Matters in Sihwa Industrial Complex, Korea. *Sustainability* **2020**, *12*, 7647. [CrossRef]
22. Brancalion, P.H.S.; Cardozo, I.V.; Camatta, A.; Aronson, J.; Rodrigues, R.R. Cultural Ecosystem Services and Popular Perceptions of the Benefits of an Ecological Restoration Project in the Brazilian Atlantic Forest. *Restor. Ecol.* **2013**, *22*, 65–71. [CrossRef]
23. Ko, H.; Son, Y. Perceptions of cultural ecosystem services in urban green spaces: A case study in Gwacheon, Republic of Korea. *Ecol. Indic.* **2018**, *91*, 299–306. [CrossRef]

24. Kovács, B.; Uchiyama, Y.; Miyake, Y.; Penker, M.; Kohsaka, R. An explorative analysis of landscape value perceptions of naturally dead and cut wood: A case study of visitors to Kaisho Forest, Aichi, Japan. *J. For. Res.* **2020**, *25*, 291–298. [CrossRef]
25. Kohsaka, R.; Tomiyoshi, M.; Saito, O.; Hashimoto, S.; Mohammend, L. Interactions of knowledge systems in shiitake mushroom production: A case study on the Noto Peninsula, Japan. *J. For. Res.* **2015**, *20*, 453–463. [CrossRef]
26. Lee, Y.; Rianti, I.P.; Park, M.S. Measuring social capital in Indonesian community forest management. *For. Sci. Technol.* **2017**, *13*, 133–141. [CrossRef]
27. Uchiyama, Y.; Kohsaka, R. Analysis of the Distribution of Forest Management Areas by the Forest Environmental Tax in Ishikawa Prefecture, Japan. *Int. J. For. Res.* **2016**, *2016*, 1–8. [CrossRef]
28. Takahashi, T.; Uchida, Y.; Ishibashi, H.; Okuda, N. Subjective Well-Being as a Potential Policy Indicator in the Context of Urbanization and Forest Restoration. *Sustainability* **2021**, *13*, 3211. [CrossRef]
29. Park, C.; Suk, S.; Choi, S. The Functional Traits of Breeding Bird Communities at Traditional Folk Villages in Korea. *Sustainability* **2020**, *12*, 9344. [CrossRef]
30. Quevedo, J.M.D.; Uchiyama, Y.; Lukman, K.M.; Kohsaka, R. How Blue Carbon Ecosystems Are Perceived by Local Communities in the Coral Triangle: Comparative and Empirical Examinations in the Philippines and Indonesia. *Sustainability* **2020**, *13*, 127. [CrossRef]
31. McCann, E.; Sullivan, S.; Erickson, D.; De Young, R. Environmental Awareness, Economic Orientation, and Farming Practices: A Comparison of Organic and Conventional Farmers. *Environ. Manag.* **1997**, *21*, 747–758. [CrossRef]
32. Quevedo, J.M.; Uchiyama, Y.; Kohsaka, R. Perceptions of local communities on mangrove forests, their services and management: Implications for Eco-DRR and blue carbon management for Eastern Samar, Philippines. *J. For. Res.* **2019**, *25*, 1–11. [CrossRef]
33. Rogan, R.; O'Connor, M.; Horwitz, P. Nowhere to hide: Awareness and perceptions of environmental change, and their influence on relationships with place. *J. Environ. Psychol.* **2005**, *25*, 147–158. [CrossRef]
34. Quevedo, J.M.D.; Uchiyama, Y.; Kohsaka, R. Linking blue carbon ecosystems with sustainable tourism: Dichotomy of urban-rural local perspectives from the Philippines. *Reg. Stud. Mar. Sci.* **2021**, *45*, 101820. [CrossRef]
35. Lukman, K.M.; Quevedo, J.M.D.; Kakinuma, K.; Uchiyama, Y.; Kohsaka, R. Indonesia Provincial Spatial Plans on mangroves in era of decentralization: Application of content analysis to 27 provinces and "blue carbon" as overlooked components. *J. For. Res.* **2019**, *24*, 341–348. [CrossRef]

Article

Relationships between Local Green Space and Human Mobility Patterns during COVID-19 for Maryland and California, USA

Seulkee Heo *, Chris C. Lim and Michelle L. Bell

School of the Environment, Yale University, New Haven, CT 06511, USA; chris.lim@yale.edu (C.C.L.); michelle.bell@yale.edu (M.L.B.)
* Correspondence: seulkee.heo@yale.edu

Received: 12 September 2020; Accepted: 9 November 2020; Published: 12 November 2020

Abstract: Human mobility is a significant factor for disease transmission. Little is known about how the environment influences mobility during a pandemic. The aim of this study was to investigate an effect of green space on mobility reductions during the early stage of the COVID-19 pandemic in Maryland and California, USA. For 230 minor civil divisions (MCD) in Maryland and 341 census county divisions (CCD) in California, we obtained mobility data from Facebook Data for Good aggregating information of people using the Facebook app on their mobile phones with location history active. The users' movement between two locations was used to calculate the number of users that traveled into an MCD (or CCD) for each day in the daytime hours between 11 March and 26 April 2020. Each MCD's (CCD's) vegetation level was estimated as the average Enhanced Vegetation Index (EVI) level for 1 January through 31 March 2020. We calculated the number of state and local parks, food retail establishments, and hospitals for each MCD (CCD). Results showed that the daily percent changes in the number of travels declined during the study period. This mobility reduction was significantly lower in Maryland MCDs with state parks (p-value = 0.045), in California CCDs with local-scale parks (p-value = 0.048). EVI showed no association with mobility in both states. This finding has implications for the potential impacts of green space on mobility under an outbreak. Future studies are needed to explore these findings and to investigate changes in health effects of green space during a pandemic.

Keywords: mobility; vegetation; green space; sustainability; social media; disease prevention

1. Introduction

The COVID-19 outbreak occurred just before the 2020 Lunar New Year in China [1] and rapidly led to a global spread. The World Health Organization (WHO, Geneva, Switzerland) declared that the new coronavirus outbreak is an international public health concern on 30 January 2020 [2] and WHO officially announced COVID-19 as a pandemic on 11 March 2020 [3]. On 16 March 2020, the US federal government announced the '30 Day to Slow' guideline in response to the pandemic. While implementation of mitigation measures varied by US state and district, many of those declarations were announced in March [4]. For example, Maryland enacted a "Prohibiting Large Gatherings and Events and Closing Senior Centers" order on 12 March 2020; a statewide "stay-at-home" order went into effect on 30 March [5]. California declared a state of emergency in March and a stay-at-home order in 19 March. Under these orders, residents were permitted to go outside "for fresh air and exercise as long as they are maintaining a safe distance from others."

Social distancing, or physical distancing, has widespread consequences, affecting the economy and individuals' behaviors in various ways [6], including significant decreases in human mobility and traffic volume around the period when various governments announced interventions (i.e., guidelines for social distancing, quarantine, and stay-at-home orders) [1,7,8]. While the definition of mobility varies among studies and disciplines, mobility analysis examines possible travel destinations and travel route based on local land use and demographics [9]. A better understanding of the mobility (e.g., destinations) of people can assist decision-making in prevention of disease transmission [10]. For this, studies showed that aggregated human mobility based on mobile phone data can assist studies assessing the spread of epidemics [11], economic consequence of the COVID-19 pandemic [6,12], and how the stay-at-home orders are effective to mitigate human mobility and thereby reduce the COVID-19 transmission [1,13,14]. Large mobility reduction was detected following the COVID-19 pandemic and specific government directives in the US and globally [15–17]. A study using mobility data from Wuhan and transmission of cases across China found that the positive relationship between human mobility and COVID-19 cases decreased after control measures [1]. A few other studies also suggested that sustained human mobility due to domestic and/or international air travel bans contributed to decreased transmission of COVID-19 cases at the early stages of the outbreak in European countries [18] and China [19]. Given the clear links between mobility and spread of the novel coronavirus [20], understanding mobility patterns is crucial to address COVID-19 outbreaks and develop policies to minimize transmission. While human mobility data have been utilized in some scientific works regarding visualizing mobility patterns [21] and economic effects of mobility changes [6], little is known about whether and how environmental factors influence human mobility under normal conditions and during the COVID-19 pandemic.

Alongside government directives for staying at home and social/physical distancing, health authorities including U.S. Centers for Disease Control and Prevention (CDC) and WHO emphasized the importance of regularly performing exercise to cope with the stress of quarantine, stay healthy, and maintain immunity [22]. Several studies argued that physical inability as a consequence of strict quarantine may be associated with increased risk of mental health outcomes as well as cardiovascular diseases, metabolic diseases, and cancer [23–26]. Thus, the COVID-19 pandemic along with other major environmental crises such as climate change shed a light on the need for better understanding of how to promote resilience or capacity of societies to deal with complex health crises [26].

Earlier work indicates that green space provides health benefits [25,27] and sustainability in cities [28]. Green space is defined as natural vegetation such as grass, bush, plants or trees and the built green structures such as parks and unstructured vegetated areas [29]. Potential pathways for the health benefits from green space include encouraging physical activities and providing direct interactions with nature [30]. Given the restrictions on the gathering of people particularly in indoor settings during COVID-19, understanding the use of green space contributes to our understanding of how green space relates to the ability of communities to cope with the stress from quarantine and pandemic, such as by playing a role as an alternative place for physical activity. A recent study in Oslo, Norway found that outdoor physical activity levels increased after the lockdown was implemented, and that the increases were highest in trails with greener and more remote areas [31]. A study conducted in the US found that the reduction in mobility to parks impacted by state-of-emergency declarations was smaller than the mobility reduction for other venues across the states [32]. Thus, green space could be an effective modifier on the effectiveness of COVID-19 mitigation measures, and such measures could indirectly impact the public health benefits of greenness.

As of April 2020, most US states have ordered nonessential businesses such as restaurant, bars, theaters, and gyms to close but the status of green spaces and open spaces such as parks have been far less consistent in many states. Green space may be one of the limited outdoor places where people seek to perform outdoor exercise during the COVID-19 crisis. The aim of this study was to investigate green space as a potential factor influencing mobility patterns during a pandemic. We hypothesized that the expected mobility decreases due to the social/physical distancing and associated policies will be lower in areas with higher local green space. Specifically, for a case study region of Maryland and California, we examined how the temporal trends in the number of people traveling among the study regions (minor civil division (MCDs) for Maryland, census county division (CCD) for California) differ by local vegetation level. The results here provide information relevant for the design and effectiveness of sheltering policies designed to mitigate a pandemic.

2. Materials and Methods

We utilized de-identified and aggregated large-scale data developed by Facebook Data for Good platform to identify population mobility trends during the COVID-19 crisis [33]. These data are called Movement Data and they aggregate information from people using the Facebook app on their mobile phones with location history turned on to show movements between two points. The computation system of this data acquires the most common location of the Facebook app user using Bing tile map Level 13 (e.g., approximately 4.9 × 4.9 km) [34] within the first time window and the most common location in the second time window for each day. The centroids of the starting and ending Bing tiles are assigned to the person's movement vector. Then, these vectors are aggregated into the administration level 4 boundary (e.g., township), which is spatially equivalent to MCDs in Maryland and CCDs in California. MCDs and CCDs are administrative county subdivisions for which the Census Bureau establish and provide subcounty statistics [35]. CCDs are equivalent geographic entities to MCDs in US states where MCDs do not exist or have been unsatisfactory for comparing statistical data [35]. The data characterized mobility trend as the percent change in the observed number of users traveling into the administrative area (MCD or CCD) for the same time window and the same day of the week compared to the baseline period at MCD (CCD) level so the mobility data indicate movement of users across regions at this spatial level. The baseline number of users moving into an MCD (CCD) was calculated as the average number of users moving for the same daytime window and day of the week between 26 February 2020 to 10 March 2020. The percent change in the number of users between a given day and its baseline is calculated as follows [33]: percent change = $(c_t - \mu_{baseline,t} / \left(\mu_{baseline,t} + \epsilon\right)$, where c is the number of users moving into a MCD (CCD) for day t, $\mu_{baseline,t}$ is the mean of the number of users traveling into the same MCD over the same time interval on the same day of the week of day t during the baseline period, and ϵ is a small value, in this case 1. The distance traveled by users for each MCD (CCD) was calculated as the distance of the movement vectors linking the centroids of the starting and end Bing tiles of all users who traveled into that MCD (CCD). We used the daytime window (8 a.m.–4 p.m.) for our analysis.

The Movement Data of our study regions were linked to the geographic information systems (GIS) data of MCD in Maryland and CCD in California provided by the US Census Bureau. We analyzed the 230 administrative areas (MCDs) out of 290 total administrative areas in the state of Maryland, US, for which the Movement Data were available, in order to examine mobility trends for the period from 10 March to 24 April 2020. For California, we analyzed 341 CCDs out of 397 CCDs, for which the Movement Data were available. Percent change in the number of users moving was not observed for some MCDs when MCDs are smaller than a Bing tile so the user's location at MCD and CCD level cannot be specified. Data are not provided for county subdivisions where the number of observed users is smaller than 10 users to protect users' privacy. County subdivisions with no observation for users' moving between pairs of county subdivisions for 70% or more of the days in the study period were excluded in the analysis. As a result, 76 MCDs in Maryland and 241 CCDs were included in our main analysis.

Vegetation level to indicate green space was estimated by the Enhanced Vegetation Index (EVI) from the Moderate Resolution Imaging Spectroradiometer (MODIS) product MOD13Q1, which is a 16-day composite image at 250-meter resolution. The EVI is an advanced version of the Normalized Different Vegetation Index (NDVI). The NDVI is calculated as near-infrared radiation minus visible radiation divided by near-infrared radiation plus visible radiation (i.e., NDVI = (NIR − RED)/(NIR + RED)). The index ranges from −1 to +1 with higher values indicating denser vegetation and −1 indicating waterbody features (NASA, 2018). While the EVI's calculation is similar to NDVI, it corrects for some distortions in the reflected light caused by the particles in the air, ground cover below the vegetation, and the saturating effects of areas with large amount of chlorophyll such as rainforests [36]. We calculated the average EVI for each study region (MCD, CCD) using the EVI pixel values within and surrounding the MCD and CCD boundary, for 1 January 2020 through 21 March 2020 to represent vegetation level in the study regions.

The MODIS Land Cover Type Product (MCD12Q1) [37] was used to estimate urbanicity of each study region. The number of 'Urban and Built-up Lands' pixels based on the University of Maryland legend and class definition was divided by the total number of pixels within the MCD or CCD boundary was calculated as the percent impervious area.

We obtained the park data in California from Esri Data & Maps [38]. These data include parks and forests at national, state, county, and local levels (e.g., city-scale parks, pocket parks, playgrounds, etc.) [38]. We obtained the Maryland State Parks data provided by the OpenStreetMap [39]. The OpenStreeMap is a global dataset including open user-generated street maps, geographical features, and built environment, which have been used for a wide range of studies [40]. This Maryland State Parks data, developed by the Baltimore County Government, is a GIS shape file and includes geographical polygons for several types of green space: state and national parks or forest, hiking trails, natural resource management areas for recreational activities (e.g., fishing, hunting, wild animal observation, etc.) and preservation of environmental resources, and wildlife management areas [41]. Data of parks at local levels in Maryland were obtained from Esri Data & Maps [38]. Hereafter, we use the term parks to refer to all these types of areas. The continuous variable for parks did not have a normal distribution. Thus, we considered a categorical variable of presence of parks within MCDs and CCDs (i.e., MCDs with state parks vs. MCDs without state parks within their boundary).

We obtained the GIS file of food retail establishments and hospitals from various sources. We obtained the data for food stores (2017–2018) and restaurants (2019) in Maryland from the Maryland Food System Map (JHSPH) developed by the Johns Hopkins Center for a Livable Future [42]. The food stores data included attributes for grocery stores, supermarkets, gas stations, and pharmacies. The GIS data of hospitals (acute, general, and special) licensed by the Maryland Department of Health and Mental Hygiene Office of Health Care Quality were obtained from Maryland's Mapping & GIS Portal [43]. Using these datasets, we calculated the sum of the number of food stores (grocery stores, supermarket, gas station, pharmacy), restaurants, pharmacies, and hospitals for each study MCD in Maryland. For California, the data on hospitals (2020) were obtained from Esri Data & Maps [38] and the data for pharmacies (2019) were obtained from OpenStreetMap [39]. The data for food retail establishments were not available for California in this study due to the lack of data for many CCDs across California.

We calculated statistics such as the first quartile (Q1), third quartile (Q3), mean, and minimum values of the daily percent changes in number of people moving between pairs of study MCDs (or CCDs) between 31 March and 24 April 2020 (i.e., after stay-at-home order) for Maryland and between 31 March and 19 April for California to characterize the mobility trends. Using linear regression analysis, we analyzed the relationships between the local vegetation level (i.e., EVI), presence of parks, and percent changes in mobility. A linear regression analysis was used for these statistics and parks, EVI, and urbanicity to examine if mobility patterns during the early stage of COVID-19 pandemic differed by green space (i.e., incorporating state parks or EVI level). We applied several different statistical models with different sets of confounders. The Q3 of mobility changes was used as a dependent variable as it represented the best normal distribution in Maryland (Supplementary Figure S1), while it was slightly skewed in California. We examined if results for green space and mobility trends were confounded by urbanicity (e.g., population, percent impervious area). We conducted regression analyses separately for each state.

3. Results

Descriptive statistics of the study regions are presented in Table 1. The higher average percent of impervious area in the MCDs in Maryland where the Facebook users' movement was observed

(32.7, SD = 29.3) compared to the MCDs for which the user's movement was not found indicate higher urbanicity level (Table 1). Average population density was higher in the MCDs where the mobility data were available. The range of EVI in the study MCDs in Maryland was narrower (0.15 to 0.29) compared to the range of EVI in the study CCDs in California (0.07 to 0.42). On the contrary, population density and percent impervious area were slightly higher in CCDs where the Facebook users' movement was not observed in California. The average of percent change in the number of users moving between pairs of the MCDs in a day and across the study MCDs in Maryland was −23.7 (SD = 21.5). The maximum reduction in number of users moved between MCDs in a day was −78.4% across all MCDs. In California, the average percent change in mobility between pairs of the CCDs in a day was −29.8 (SD = 25.7) along with the maximum percent change in mobility of −94.6%. The Q1, Q3, and median of percent changes in the number of travelling users indicate that mobility declined during the study period in most regions in both Maryland and California, although mobility did increase for some regions. The distance travelled between the study regions gradually decreased during the study period (Supplementary Figure S3).

Figure 1 represents the average daily percent changes in the number of users traveling into the study areas during the COVID-19 pandemic in Maryland and California. The trend showed a decreasing pattern from the beginning of the study period until the end of March and remained constantly at a low level until the end of the study period and the decrease in mobility was the highest in MCDs with the lowest level of EVI (i.e., <0.21) in Maryland. The decrease in mobility was the lowest in CCDs with the medium level of EVI (i.e., 0.24–0.29) in California.

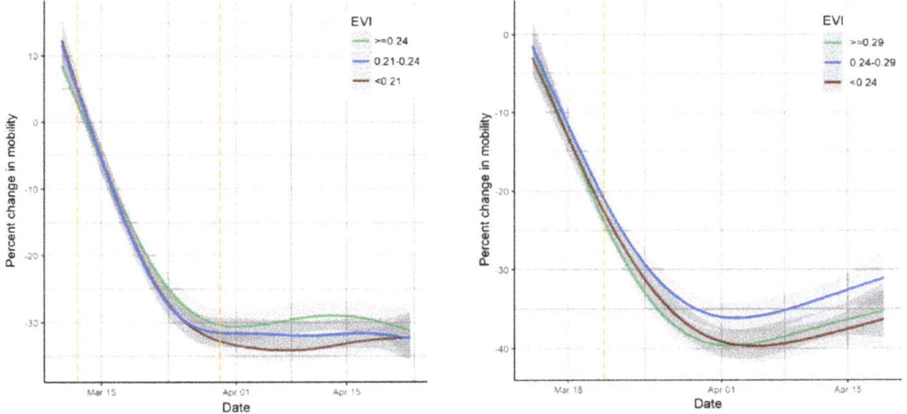

Figure 1. Average daily percent changes in mobility among study areas by EVI level in Maryland (**left**) and California (**right**) during the COVID-19 pandemic. Percent change in mobility is the percent change in the number of users traveling in the daytime (8 a.m.–4 p.m.) in a given day compared to the same time and the same day of the week in the reference period (26 February–10 March). Solid line: LOESS smoothing line; grey area: 95% confidence interval of LOESS line; yellow dotted lines for Maryland: declaration of state of emergency and the following stay-at-home order; yellow dotted line for California: stay-at-home order.

Table 1. Descriptive statistics of vegetation level and mobility trends during the COVID-19 pandemic by minor civil division (MCD) in Maryland (11 March–24 April 2020) and census country division (CCD) in California (11 March–19 April 2020).

Variable	MCDs Where Mobility Was Observed					MCDs Where Mobility Was Not Observed				
	Mean (SD)	Min-Max	Q1	Q3	Median	Mean (SD)	Min-Max	Q1	Q3	Median
Maryland		$n = 76$					$n = 154$			
EVI (range −1 to 1)	0.22 (0.03)	0.15–0.29	0.19	0.23	0.22	0.25 (0.04)	0.05–0.31	0.23	0.27	0.25
Percent of impervious area (%)	32.7 (29.3)	0.0–92.3	7.2	61.8	19.0	1.3 (3.5)	0.0–33.3	0.0	1.1	0.1
Population (persons)	61,080 (74,559)	161–566,200	25,810	84,830	44,920	8892 (8860)	436–39,470	2224	13,030	6319
Population density (persons/km^2)	805.3 (901.9)	0.4–6062.0	218.4	991.4	563	90.5 (100.6)	3.1–600.8	21.3	132.0	58.2
Number of food retail and hospitals *	187 (321)	2–2712	58	206	126					
Percent change in mobility (%)	−23.7 (21.5)	−78.4–120.7	−38.9	−10.8	−26.3					
Travel distance (km) †	7.3 (3.4)	3.8–23.2	5.3	8.4	6.8					
Presence of parks *	Number	%				Number	%			
Yes	65	85.6				123	79.9			
No	11	14.4				31	20.1			
California		$n = 241$					$n = 156$			
EVI (range −1 to 1)	0.26 (0.07)	0.07–0.42	0.22	0.31	0.27	0.27 (0.07)	0.07–0.31	0.22	0.32	0.27
Percent of impervious area (%)	10.4 (19.7)	0.0–100.0	0.2	8.3	1.3	12.7 (23.7)	0.0–100.0	0.4	10.7	2.2
Population (persons)	110,587 (294,124)	741–2,457,972	7026	74,510	20,785	67,965 (174,581)	262–1,664,311	5531	59,500	12,750
Population density (persons/km^2)	247.5 (584.5)	0.1–5136.7	5.6	166.0	34.1	303.7 (680.3)	0.1–4951.0	13.9	221.7	49.3
Number of hospitals and pharmacies	6.1 (19.0)	0.0–190.0	0.0	5.0	1.0	3.4 (9.7)	0.0–98.0	0.0	3.0	1.0
Percent change in mobility (%)	−29.8 (25.7)	−94.6–377.9	−47.2	−16.0	34.3					
Travel distance (km) †										
Presence of parks *	Number	%				Number	%			
Yes	216	89.6				118	75.6			
No	25	10.4				38	24.4			

* Food retail establishments and hospitals refer to grocery stores, supermarkets, gas stations, pharmacies, restaurants, and hospitals. The 'presence of parks' variable refers to national parks and forest, state parks and forests, hiking trails, and local-scale parks. † The average distance between the centroids of spatial grid cells (Bing tile) the users traveled between for each day during the study period. EVI = Enhanced Vegetation Index.

Figure 2 shows the descriptive statistics (Q1, Q3, median, and average) of daily percent changes in mobility and EVI values at the MCD level (or CCD level) of the study regions. The scatter plots showed that the regions with high EVI values may have lower reduction in their mobility trend during the study period in Maryland and California. However, the correlation coefficients for the EVI in Maryland were 0.11, −0.01, 0.08, and 0.05 for the Q3, Q1, mean, and median of mobility changes, respectively, indicating no significant correlations. Similarly, for California, no significant correlations were observed for EVI and statistics of mobility changes (0.07, 0.12, 0.08, and 0.08 for the Q3, Q1, mean, and median of mobility changes, respectively).

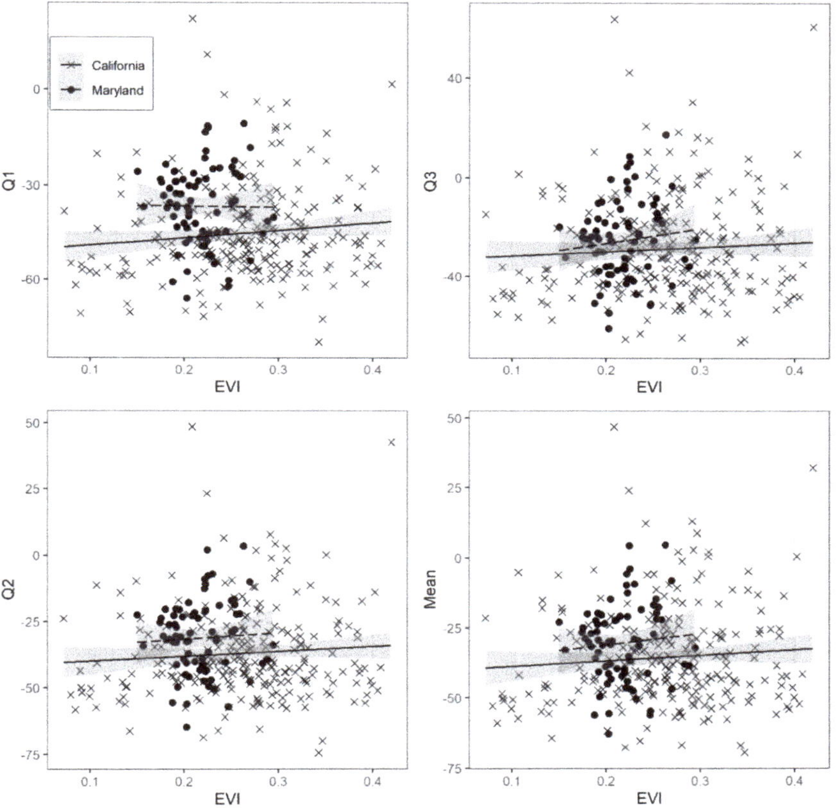

Figure 2. Scatter plots of statistics of percent changes in mobility and EVI in the study areas in Maryland (n = 76) and California (n = 241). The grey area is 95% confidence intervals for linear regression lines.

Figure 3 presents the locations of parks and forest and the geographical patterns of mobility changes (Q3) and EVI. On average, the sizes of parks and forests as well as the size of county subdivisions were larger for California than Maryland. The movement of users was mostly observed in the central areas of Maryland including MCDs adjacent to Baltimore, Maryland. The geographical pattern of EVI showed a relatively particular pattern with higher EVI values for central western parts in Maryland and western parks of California, while the geographical patterns of mobility changes appeared to be random in study regions in both states.

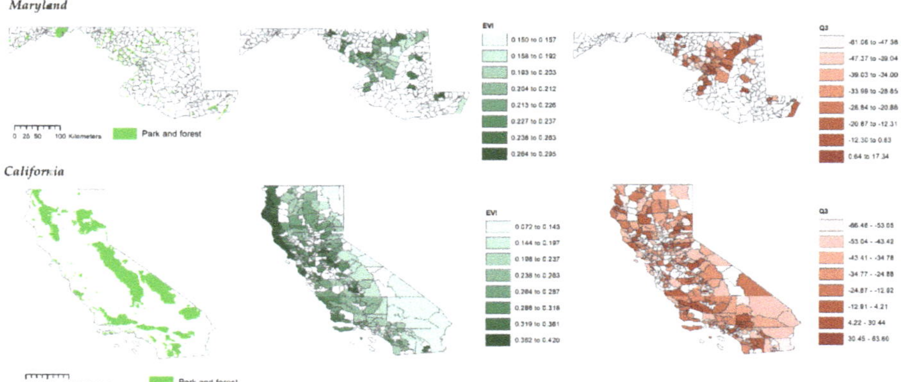

Figure 3. Location of parks and forest and the spatial patterns of EVI and the statistics (Q1, mean, Q3) of percent changes in the number of users moving into each subdivision in Maryland (top) and California (bottom) during the study period (11 March–26 April 2020). Blank area: County subdivisions (MCDs, CCDs) where users' movement data were unavailable.

Figure 4 presents the mobility change and EVI for MCD (CCD) groups with and without parks. MCDs with parks in Maryland showed slightly lower reduction in mobility compared to MCDs without parks. In California, reduction in mobility was relatively lower in CCDs with parks compared to CCDs without parks. EVI was lower in MCDs and CCDs with parks in Maryland and California.

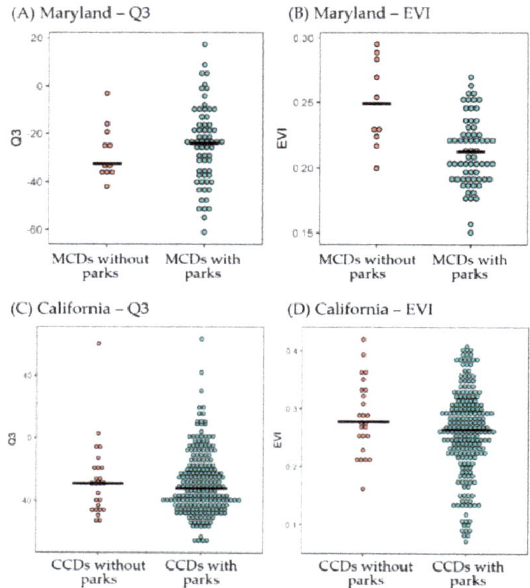

Figure 4. Statistics of percent changes in mobility (Q3) and EVI for MCD (CCD) groups with and without parks in Maryland and California. Solid lines are median.

Regression coefficients of EVI from the linear regression analysis are presented in Table 2. Although MCDs (CCDs) with high EVI values tended to have lower reduction in their mobility (Figure 2), EVI at the MCD (CCD) level in Maryland and California were not significantly associated with mobility changes (Q3) during the study period in any model.

Table 2. Regression coefficients of EVI for the relationship with mobility changes (Q3) in the study county subdivisions (MCD, CCD) in Maryland and California.

Region	Model 1			Model 2		
	Beta	(95% CI)	*p*-Value	Beta	(95% CI)	*p*-Value
Maryland (*n* = 75)	−1.06	(−2.49, 0.38)	0.154	−1.31	(−2.63, 0.02)	0.067
California (*n* = 241)	0.02	(−9.02, 9.74)	0.933	0.13	(−0.23 0.50)	0.473

Note: EVI = Enhanced Vegetation Index. Model 1 was adjusted for presence of parks (all types), log population density, and number of food retail establishments and hospitals; Model 2 was adjusted for presence of parks (all types), percent impervious area, and number of food retail establishments and hospitals.

Results of the linear regression analysis for presence of parks are shown in Table 3. Presence of state parks in MCD boundary in Maryland was significantly associated with lower reduction in mobility (7.62, 95% CI: 0.28, 14.97) in Model 1, whereas Model 2 and Model 3 did not show significantly lower reduction in mobility. When percent impervious area was included as adjustment instead of population density, having state parks or EVI did not show significant effects on mobility changes. In the third model incorporating the variable of presence of state parks, number of food retail establishments and hospitals, and population density, presence of state parks showed significant effects on mobility changes at a 0.10 significance level. All types of parks and local-scale parks showed no significant relationship with mobility changes in Maryland during the study period. On the other hand, presence of local-scale parks in CCDs in California showed significantly lower reduction in mobility in Model 1 (7.03, 95% CI: 0.12, 13.94) and Model 3 (7.02, 95% CI: 0.13, 13.92). Mobility reduction tended to be lower in CCDs with any type of parks or state or national parks but the results were not significant in California.

Table 3. Results of linear regression analysis on the mobility changes (Q3 and presence of parks in the study county subdivisions (MCD, CCD) in Maryland and California.

Variable	Model 1			Model 2			Model 3		
	Beta	(95% CI)	*p*-Value	Beta	(95% CI)	*p*-Value	Beta	(95% CI)	*p*-Value
Maryland (*n* = 75)									
Presence of parks by type (ref = having no park)									
Total	7.31	(−4.97, 1.96)	0.248	1.57	(−9.96, 13.10)	0.790	7.00	(−4.61, 18.60)	0.242
State and National	7.62	(0.27, 14.97)	0.045	5.56	(−1.87, 13.00)	0.145	6.04	(−1.05, 13.13)	0.100
Local	5.79	(−5.21, 16.80)	0.306	0.057	(−9.52, 10.65)	0.913	5.71	(−4.87, 16.29)	0.294
California (*n* = 241)									
Presence of parks by type (ref = having no park)									
Total	0.36	(−9.02, 9.75)	0.939	−0.04	(−9.71, 8.87)	0.930	0.33	(−9.00, 9.66)	0.945
State and National	0.22	(−6.20, 6.65)	0.946	0.38	(−5.91, 6.68)	0.906	0.20	(−6.19, 6.58)	0.952
Local	7.03	(0.12, 13.94)	0.048	5.17	(−1.01, 11.36)	0.102	7.02	(0.13, 13.92)	0.047

Note: Model 1 was adjusted for number of food retail establishments and hospitals, population density, and EVI; Model 2 was adjusted for number of food retail establishments and hospitals, percent impervious area, and EVI; Model 3 was adjusted for number of food retail establishments and hospitals and log population density.

4. Discussion

We examined if the amount of green space such as parks and vegetation level are associated with human mobility under pandemic mitigation policies due to COVID-19 in Maryland and California. A novel result of this study is that the decline of mobility during the COVID-19 pandemic appeared to be lower in regions with green space such as state parks in Maryland and local parks in California. This may imply that people sought green spaces or beach areas to cope with the stress of the pandemic and to perform outdoor activities with social distancing. Further, our results demonstrated that types and scales of parks (e.g., state, local) may impact the effects of parks on mobility changes in different US states potentially due to differences in size of administrative regions (e.g., county subdivision), population density, land cover, or scales of green space (e.g., parks, forest, beach). Our study does not provide data for an association between mobility and COVID-19 transmission. Nonetheless, the potential impact of green space on mobility during a pandemic shown in our study and other recent literature [7,31,32] implies the need for future design and planning for green space and open spaces where disease control measures such as physical distancing can be performed during an outbreak [31,44,45]. Our results suggest that different plans for sheltering in place in relation to local built environment at regional levels are required. Furthermore, this study is in the scope of sustainable development for resilience and preparedness with nature against future pandemics [46]. Currently, there is an urgent need for studying how mental health consequences can be mitigated in a pandemic [24]. Previous work suggested that a long quarantine duration can be a major stressor causing emotional distress and increased risk of psychiatric illness, unhealthy behaviors, and noncompliance with mandate public health guidelines [47]. A study conducted in Italy provided evidence that the COVID-19 pandemic was significantly associated with increased risks of developing depression, anxiety, and sleeping disorders during lockdown [48]. Given the impact of the pandemic on mental health, and the benefits of green space to mental health, understanding the relationship between the pandemic and green space is paramount.

Many studies have utilized the vegetation index to assess amount of local green space in relation to provision of green space [49,50] and health effects of green space [51]. Given the negative correlation between EVI and urbanization (e.g., population density, impervious areas) in our study, the differences in mobility reduction by EVI levels reflect that people traveled less to more populated areas for safety. This was seen for the graph of daily mobility changes by EVI levels in California. Our data also showed that the presence of parks did not corresponded with higher vegetation level at the MCD (CCD) level in Maryland and California, which indicates that parks were more located in urbanized areas. To control for the effect of urbanization, we examined two types of urbanization indicators (i.e., population density, percent impervious area) in our statistical models. Our results found that the presence of parks and other types of natural areas for recreational activities (e.g., fishing, hunting) was more significantly associated with mobility patterns than was vegetation level. This was particularly due to the characteristics of land cover in CCDs in California where state and national parks were located. These large-scale parks were located in CCDs in central eastern parts of California, where there are also deserts such as the Mojave Desert bounded by the Tehachapi Mountains, San Gabriel and San Bernardino Mountains. These areas are also bounded by Arizona and Nevada where the land cover is largely desert at the borders with California. Due to these characteristics, EVI tended to be lower in these areas despite the presence of state and national parks. On the other hand, lower effects of EVI could be related with the nature of the vegetation index as it does not fully represent the quality of greenness or the actual accessibility (e.g., presence of entry points for people) to green space, although the vegetation index has been used extensively to measure greenness [52]. Study results for the best greenness metrics for assessing health effects are inconsistent. Thus, we suggest that future studies on green space and health consider types and volume of green space in addition to the vegetation index to better understand the functions of green space for outdoor activities and in relation to health, and that research on the characteristics of green space that are most relevant for health be conducted.

Studies suggested that transmission of COVID-19 can be mitigated with further local control measures including suspending public transport, closing entertainment venues, and banning public gatherings in addition to travel bans [1,53]. In response to the pandemic, the Maryland Department of Natural Resources postponed and/or canceled programs and events and closed areas in parks where the public may congregate, such as visitor centers, administrative buildings, and shelters [54]. Several state parks and golf courses were closed to the public during our study period. Gatherings of more than 10 persons were prohibited, and state residents were encouraged to keep distance when outside, not participate in team sports, and avoid touching surfaces that may be handled by others (e.g., playground equipment and benches). Access to most parks in Maryland as of April 2020 was not restricted due to COVID-19. Parks and trails remained open for activities such as hiking, biking, or walking, although state park beach areas were closed on 30 March [55]. Thus, we assumed that the closure of some parks due to COVID-19 did not affect our analysis on mobility and green space, although uncertainty remains. Similarly, California Department of Parks and Recreation announced its first order for suspension of tours to some California State Parks in 14 March in an effort to protect public health from COVID-19 and announced additional temporary full closure of state parks in 3 April [56]. Since then, access to California State Parks has continuously changed according to the severity of disease spread and compliance with state and local public health ordinances of communities. It is important to note that having different regulations for access to green space hinders direct comparisons of the relationships between green space and human mobility among different US states. This may justify analysis for individual states or administrative units with similar scales.

While our study focused on the increases in visits to green space at an early stage of the pandemic, which was also identified in other literature [7,31,32], COVID-19 has deteriorated the ability of people to utilize urban green space resources (e.g., parks, playgrounds) in many communities [57]. Hall et al. suggested that efforts should be made to assess the lasting effects of pandemics on physical activities along with closures of open green spaces [57]. In addition to such call for research, our empirical results on the relationships between green space and human mobility suggest the need for further understanding and study directions about the way people use green space. Although crowding in urban parks is likely to contribute to transmission of COVID-19, timely data for the number of visitors in green space are not readily available. Further, it is unknown if and how people perform safety measures such as wearing masks and practicing physical distancing in public green spaces. While numerous studies have provided evidence of health benefits of green space on psychological and physical health through stress relief, enhanced physical activities, and social cohesion [25,27], there is a lack of information on which benefits occur when people interact with green space during the pandemic with social distancing policies, and how these relate to the health detriments of increased disease transmission. It is unclear if the frequency or purposes of visiting green space have changed due to the COVID-19 pandemic and, if so, how those changes and resultant health effects differ by subpopulation (e.g., race/ethnicity, socioeconomic status). Future studies will need to address the complex relationships among COVID-19, greens space, and health effects in order to better understand how the pandemic affects our interaction with green space and its health impacts.

Our study has several strengths. To our best of our knowledge, this study is the first to examine the effect of amount of green space on human mobility during the COVID-19 pandemic in Maryland and California. We used novel social media app-based mobility data and considered the density of basic social assets (e.g., number of food retail establishments and hospitals) within county subdivisions (MCDs, CCDs) as an indirect indicator of people's travel outside county subdivisions for basic services (e.g., grocery shopping, health care services). Our results provide timely information relevant for the policies of control measures of a pandemic in relation to green spaces.

This study has some limitations. The mobility identified based on app-based location services can only be observed when the app is active and is thereby may be affected by situations for which people are likely to use a phone (e.g., searching directions, connecting with friends, posting photos). We were not able to consider user's movement within a Bing tile. We could not consider all county subdivisions in Maryland and California due to the lack of mobility data in some subregions based on the privacy protection and lack of Facebook app users. Facebook Data for Good provides daily mobility changes, only since March 2020 with February as the reference period, which prevents comparisons of mobility patterns across years (e.g., the same day one year before) and prevents analysis of mobility patterns during the pre-pandemic period. As a result, we could not disentangle mobility changes due to the state-of-emergency declaration from a potential seasonal pattern of mobility though February–April. A recent study argued that seasonality rather than the COVID-19 pandemic caused the reported increases in park visitations in Google Mobility Reports for some Western US counties [58]. We note that our study area likely shares the same seasonality of mobility across county subdivisions, therefore, the potential impact of seasonality does not likely explain differences in mobility across study subregions. We also note that our main question was the impact of green space on mobility patterns during the COVID-19 pandemic rather than the impact of the state-of-emergency declaration or specific policies or guidance, which may vary locally, on mobility patterns during the lockdown period. Another limitation is that we could not consider Baltimore city in our regression analysis for Maryland as it was an extreme outlier in terms of population size and number of food retail establishment and hospitals compared to other MCDs, which affected our assumption for the linear relationships for mobility changes and the covariates in the model. Although we used commonly applied datasets for food retail establishments in Maryland, these measures do not include every type of food supply store (e.g., farmer's markets, food trucks). Therefore, there is an uncertainty regarding weather such results would be robust when the choices of types of food retail or health care facilities are different. In addition, food retail data were not available for California in this study. Thus, a cautious interpretation is required of the study results and their comparisons between the study states. We note that our findings of the relationship between presence of parks and mobility changes during the study period may have limited generalizability beyond the sub-urban or peripheral urban areas. This leaves a question for the impact of urbanicity on the use of green space and requires future studies comparing highly urbanized cities in the US. Only Maryland and California were examined in our analysis. The Facebook Data for Good has gradually published open data for mobility for more expended regions since February 2020. However, those updated mobility data have a discrepancy for observation period with previous datasets. We considered Maryland and California, which had the same reference period for daily estimating mobility changes across the US.

Due to the unprecedented nature of the pandemic and the critical need for scientific evidence, many research studies related to COVID-19 have been conducted in an urgent manner in spite of methodological challenges, which leads to some degrees of uncertainty. For the sake of public health against the novel impact of COVID-19, our study examined a unique hypothesis for the relationship between green space, COVID-19, and mobility by combining social media data and satellite remote sensing technology. However, future work is needed to confirm the findings presented here and to investigate relevant questions. Here we discuss suggestions for future studies. First, higher resolution of mobility data would aid understanding of neighborhood-level mobility patterns and their relationships with green space. Individual-level data could illuminate differences in patterns by population characteristics, such as by socioeconomic status and race/ethnicity, which is important given the higher health burden faced by these groups. However, ethical issues of privacy should be considered and continuous efforts to find the best methodological approach for mobility data with lower spatial resolution and with individual-level data. Discussions on finding a scientifically and socially acceptable trade-off between privacy and scientific data needs might support future research. Second, there are various types of mobility data available. While Facebook Data for Good only provides mobility data of Facebook users, mobility data are obtainable through mobile phone networks (i.e., cellular networks). Mobile phone networks are composed of geographic zones (called 'cell') around a phone tower and each mobile phone can be located by identifying the geographical location and the associated cell of its transmitting phone tower [59]. This type of mobility data will incorporate any people who use mobile phone services for a given geographical zone. Facebook's mobility data still have strengths of researcher-friendly pre-generated outcomes by the Facebook Data for Good team, whereas the data based on mobile phone networks from phone towers may require the researcher's own computation logics and detailed understanding of such data. Another type of mobility data is specifically available for "point-of-interest (POI)." Mobility to POI data can be considered in assessing the mobility patterns to each destination of green space including parks and they may be less sensitive to challenges from low spatial resolutions of mobility estimation. Some datasets are freely shared for research through an existing collaborative consortium between the data company and research organizations. Considering this type of datasets based on POI would be helpful to understand the health effects of mobility to green space. Third, there are inconsistencies in the definition of green space among different countries and disciplines, and while various datasets on green space provide information, none portray the rich characteristics of heterogeneity in green space such as different types of vegetation, park access, park features, etc. Differences in scales, features, and quality in green and open space may result in difficulties in comparing study results among different regions and could obscure important relationships between mobility and green space. It is imperative to understand sizes of geographic divisions, environmental characteristics (e.g., land cover), green space features (e.g., parking availability), and urbanization in study regions to investigate the effects of green space on human mobility. Fourth, more information should be produced for human behavior patterns for using green space. The pattern of using green space (e.g., purposes of visiting) would vary among communities by urbanicity, culture, or safety level, but less is known for these patterns. Studies, possibly with survey of local residents, would be helpful for future research. In spite of these limitations, our work aims to inform future studies of urban sustainability and public health in relation to green space under the threats of global pandemic.

5. Conclusions

The pandemic has likely changed our relationship with green space, which has numerous established public health and societal benefits, and the nature of this change is not fully understood. Scientific evidence is needed regarding travel to green space, an important environmental determinant of human health, under normal conditions and pandemics. We investigated the effect of green space on human mobility patterns during the COVID-19 pandemic using large-scale mobility data from social media; our findings imply potential increases in usage of green space in Maryland and California, USA, particularly parks and environment for recreational activities when other essential social activities were prohibited or discouraged due to the control measures of COVID-19. Results suggest that understanding environmental factors associated with mobility changes during a pandemic can aid decision-makers with preparing preventive measures against public health burdens caused by pandemics. We urge future studies to explore these findings, expand relevant data sources, investigate methodologies to estimate the health benefits of green space impeded by a pandemic and the most vulnerable persons to such damages.

Supplementary Materials: The following are available online at http://www.mdpi.com/2071-1050/12/22/9401/s1, Figure S1: Histograms of the statistics of the daily percent changes in mobility in Maryland, Figure S2: Histograms of the statistics of the daily percent changes in mobility in California, Figure S3: Trend in the distance traveled by users during the study period, Figure S4: Pair-wise scatter plots for EVI and covariates in Maryland, Figure S5: Pair-wise scatter plots for EVI and covariates in California, Table S1: Regression coefficients of covariates in the statistical models including parks (all types), EVI, retail and hospitals, population density, and percent impervious area.

Author Contributions: Conceptualization, S.H.; methodology, S.H.; software, S.H.; validation, S.H. and M.L.B.; formal analysis, S.H.; investigation, S.H.; resources, S.H.; data curation, S.H.; writing—original draft preparation, S.H. and C.C.L.; writing—review and editing, S.H. and M.L.B.; visualization, S.H. and M.L.B.; supervision, M.L.B.; project administration, M.L.B.; funding acquisition, M.L.B. All authors have read and agreed to the published version of the manuscript.

Funding: This research was funded by the U.S. Environmental Protection Agency, grant number RD835871.

Acknowledgments: This publication was developed under Assistance Agreement No. RD835871 awarded by the U.S. Environmental Protection Agency to Yale University. It has not been formally reviewed by EPA. EPA does not endorse any products or commercial services mentioned in this publication. The content is solely the responsibility of the authors and does not necessarily represent the official views of the EPA. Authors appreciate Facebook for providing Facebook Disease Prevention Maps (https://dataforgood.fb.com/tools/disease-prevention-maps/).

Conflicts of Interest: The authors declare no conflict of interest.

References

1. Kraemer, M.; Yang, C.; Gutierrez, B.; Wu, C.; Klein, B.; Pigott, D.; Open COVID-19 Data Working Group; Plessis, L.; Faria, N.; Li, R.; et al. The effect of human mobility and control measures on the COVID-19 epidemic in China. *Science* **2020**, *21*, 1–9. [CrossRef]
2. Chen, P.; Mao, L.; Nassis, G.P.; Harmer, P.; Ainsworth, B.E.; Li, F. The need to maintain regular physical activity while taking precautions. *J. Sport Health Sci.* **2020**, *9*, 103–104. [CrossRef]

3. WHO. WHO Announces COVID-19 Outbreak a Pandemic. Available online: http://www.euro.who.int/en/health-topics/health-emergencies/coronavirus-covid-19/news/news/2020/3/who-announces-covid-19-outbreak-a-pandemic (accessed on 10 April 2020).
4. Son, J.-Y.; Fong, K.C.; Heo, S.; Kim, H.; Lim, C.C.; Bell, M.L. Reductions in mortality resulting from reduced air pollution levels due to COVID-19 mitigation measures. *Sci. Total Environ.* **2020**, *744*, 141012. [CrossRef]
5. Lawrence, J.; Hogan, J.G. Order of the Governor of the State of Maryland. 2020. Available online: https://governor.maryland.gov/wp-content/uploads/2020/03/Gatherings-FOURTH-AMENDED-3.30.20.pdf (accessed on 20 April 2020).
6. Bonaccorsi, G.; Pierri, F.; Cinelli, M.; Porcelli, F.; Galeazzi, A.; Flori, A.; Schmidth, A.L.; Valensise, C.M.; Scala, A.; Quattrociocchi, W.; et al. Evidence of economic segregation from mobility lockdown during COVID-19 epidemic. *Proc. Natl. Acad. Sci. USA* **2020**. [CrossRef]
7. Google COVID-19 Community Mobility Reports. Available online: https://www.google.com/covid19/mobility/ (accessed on 10 April 2020).
8. Jin, K. Data for Good: New Tools to Help Health Researchers Track and Combat COVID-19. Available online: https://about.fb.com/news/2020/04/data-for-good/ (accessed on 10 April 2020).
9. Toch, E.; Lerner, B.; Ben-Zion, E.; Ben-Gal, I. Analyzing large-scale human mobility data: A survey of machine learning methods and applications. *Knowl. Inf. Syst.* **2019**, *58*, 501–523. [CrossRef]
10. Zhou, C.; Su, F.; Pei, T.; Zhang, A.; Du, Y.; Luo, B.; Cao, Z.; Wang, J.; Yuan, W.; Zhu, Y. COVID-19: Challenges to GIS with big data. *Geogr. Sustain.* **2020**, *1*, 77–87. [CrossRef]
11. Oliver, N.; Lepri, B.; Sterly, H.; Lambiotte, R.; Deletaille, S.; De Nadai, M.; Letouzé, E.; Salah, A.A.; Benjamins, R.; Cattuto, C. Mobile phone data for informing public health actions across the COVID-19 pandemic life cycle. *Sci. Adv.* **2020**. [CrossRef]
12. Bonaccorsi, G.; Pierri, F.; Cinelli, M.; Flori, A.; Galeazzi, A.; Porcelli, F.; Schmidt, A.L.; Valensise, C.M.; Scala, A.; Quattrociocchi, W. Economic and social consequences of human mobility restrictions under COVID-19. *Proc. Natl. Acad. Sci. USA* **2020**, *117*, 15530–15535. [CrossRef]
13. Layer, R.M.; Fosdick, B.; Larremore, D.B.; Bradshaw, M.; Doherty, P. Case Study: Using Facebook Data to Monitor Adherence to Stay-at-home Orders in Colorado and Utah. *medRxiv* **2020**. [CrossRef]
14. Chang, M.-C.; Kahn, R.; Li, Y.-A.; Lee, C.-S.; Buckee, C.O.; Chang, H.-H. Variation in human mobility and its impact on the risk of future COVID-19 outbreaks in Taiwan. *medRxiv* **2020**. [CrossRef]
15. Warren, M.S.; Skillman, S.W. Mobility Changes in Response to COVID-19. *arXiv* **2020**, arXiv:2003.14228.
16. Queiroz, L.; Ferraz, A.; Melo, J.L.; Barboza, G.; Urbanski, A.H.; Nicolau, A.; Oliva, S.; Nakaya, H. Large-scale assessment of human mobility during COVID-19 outbreak. *arXiv* **2020**, arXiv:2003.14228.
17. Lasry, A.; Kidder, D.; Hast, M.; Poovey, J.; Sunshine, G.; Zviedrite, N.; Ahmed, F.; Ethier, K.A. Timing of Cmmunity Mitigation and Changes in Reported COVID-19 and Community Mobility—Four US Metropolitan Areas, 26 February–1 April 2020. Available online: https://www.cdc.gov/mmwr/volumes/69/wr/mm6915e2.htm (accessed on 11 November 2020).
18. Linka, K.; Peirlinck, M.; Sahli Costabal, F.; Kuhl, E. Outbreak dynamics of COVID-19 in Europe and the effect of travel restrictions. *Comput. Methods Biomech. Biomed. Eng.* **2020**, *23*, 710–717. [CrossRef] [PubMed]
19. Kucharski, A.J.; Russell, T.W.; Diamond, C.; Liu, Y.; Edmunds, J.; Funk, S.; Eggo, R.M.; Sun, F.; Jit, M.; Munday, J.D. Early dynamics of transmission and control of COVID-19: A mathematical modelling study. *Lancet Infect. Dis.* **2020**, *20*, 553–558. [CrossRef]
20. WHO. Report of the WHO-China Joint Mission on Coronavirus Disease 2019 (COVID-19). 2020. Available online: https://www.who.int/docs/default-source/coronaviruse/who-china-joint-mission-on-covid-19-final-report.pdf (accessed on 19 April 2020).
21. COVID-19 Mobility Data Network. Available online: https://www.covid19mobility.org/ (accessed on 18 January 2020).
22. Centers for Disease Control and Prevention Coronavirus Disease 2019 (COVID-19): Daily Life and Coping. Available online: https://www.cdc.gov/coronavirus/2019-ncov/daily-life-coping/index.html (accessed on 21 January 2020).
23. Lippi, G.; Henry, B.M.; Sanchis-Gomar, F. Physical inactivity and cardiovascular disease at the time of coronavirus disease 2019 (COVID-19). *Eur. J. Prev. Cardiol.* **2020**, *27*, 906–908. [CrossRef] [PubMed]

24. Holmes, E.A.; O'Connor, R.C.; Perry, V.H.; Tracey, I.; Wessely, S.; Arseneault, L.; Ballard, C.; Christensen, H.; Silver, R.C.; Everall, I. Multidisciplinary research priorities for the COVID-19 pandemic: A call for action for mental health science. *Lancet Psychiatry* **2020**, *7*, 547–560. [CrossRef]
25. Twohig-Bennett, C.; Jones, A. The health benefits of the great outdoors: A systematic review and meta-analysis of greenspace exposure and health outcomes. *Environ. Res.* **2018**, *166*, 628–637. [CrossRef]
26. Colding, J.; Giusti, M.; Haga, A.; Wallhagen, M.; Barthel, S. Enabling relationships with nature in cities. *Sustainability* **2020**, *12*, 4394. [CrossRef]
27. Richardson, E.A.; Pearce, J.; Mitchell, R.; Kingham, S. Role of physical activity in the relationship between urban green space and health. *Public Health* **2013**, *127*, 318–324. [CrossRef]
28. Lee, Y.C.; Kim, K.H. Attitudes of citizens towards urban parks and green spaces for urban sustainability: The case of Gyeongsan City, Republic of Korea. *Sustainability* **2015**, *7*, 8240–8254. [CrossRef]
29. Lachowycz, K.; Jones, A.P. Towards a better understanding of the relationship between greenspace and health: Development of a theoretical framework. *Landsc. Urban Plan.* **2013**, *118*, 62–69. [CrossRef]
30. Markevych, I.; Schoierer, J.; Hartig, T.; Chudnovsky, A.; Hystad, P.; Dzhambov, A.M.; de Vries, S.; Triguero-Mas, M.; Brauer, M.; Nieuwenhuijsen, M.J.; et al. Exploring pathways linking greenspace to health: Theoretical and methodological guidance. *Environ. Res.* **2017**, *158*, 301–317. [CrossRef] [PubMed]
31. Venter, Z.; Barton, D.; Gundersen, V.; Figari, H.; Nowell, M. Urban nature in a time of crisis: Recreational use of green space increases during the COVID-19 outbreak in Oslo, Norway. *Environ. Res. Lett.* **2020**, *10*, 104075. [CrossRef]
32. Wellenius, G.A.; Vispute, S.; Espinosa, V.; Fabrikant, A.; Tsai, T.C.; Hennessy, J.; Williams, B.; Gadepalli, K.; Boulange, A.; Pearce, A. Impacts of state-level policies on social distancing in the united states using aggregated mobility data during the covid-19 pandemic. *arXiv* **2020**, arXiv:2004.10172.
33. Maas, P.; Gros, A.; McGorman, L.; Alex Dow, P.; Iyer, S.; Park, W.; Nayak, C. Facebook Disaster Maps: Aggregate Insights for Crisis Response & Recovery. Available online: https://research.fb.com/wp-content/uploads/2019/04/iscram19_camera_ready.pdf (accessed on 16 April 2020).
34. Schwartz, J. Bing Maps Tile System. Available online: https://docs.microsoft.com/en-us/bingmaps/articles/bing-maps-tile-system (accessed on 23 March 2020).
35. Bureau of the Census. *Census County Divisions (CCDs) and Equivalent Entities for the 2020 Census-Final Criteria*; Bureau of the Census: Suitland, MD, USA, 2018.
36. Fensholt, R. Earth observation of vegetation status in the Sahelian and Sudanian West Africa: Comparison of Terra MODIS and NOAA AVHRR satellite data. *Int. J. Remote Sens.* **2004**, *25*, 1641–1659. [CrossRef]
37. Sulla-Menashe, D.; Friedl, M. *User Guide to Collection 6 MODIS Land Cover (MCD12Q1 and MCD12C1) Product*; USGS: Reston, VA, USA, 2018.
38. Esri Data & Maps USA Parks. Available online: https://www.arcgis.com/home/item.html?id=578968f975774d3fab79fe56c8c90941 (accessed on 11 November 2020).
39. Haklay, M.; Weber, P. Openstreetmap: User-generated street maps. *IEEE Pervasive Comput.* **2008**, *7*, 12–18. [CrossRef]
40. Ferster, C.; Fischer, J.; Manaugh, K.; Nelson, T.; Winters, M. Using OpenStreetMap to inventory bicycle infrastructure: A comparison with open data from cities. *Int. J. Sustain. Transp.* **2020**, *14*, 64–73. [CrossRef]
41. ElliottPlack The OpenStreetMap: Maryland State Parks. Available online: https://wiki.openstreetmap.org/wiki/Maryland_State_Parks (accessed on 5 September 2020).
42. Johns Hopkins Center for a Livable Future; USDA SNAP Retail Locator. USA Maryland Food Stores 2017–2018. Available online: http://data-clf.opendata.arcgis.com/datasets/c4a2bd61eaac4425b3e2e9c40735a7ae_218 (accessed on 5 September 2020).
43. Government, M. Maryland Hospitals. Available online: https://data.imap.maryland.gov/datasets/cc9954d3db7c4da18322f533a8c25ea3_0 (accessed on 5 September 2020).
44. Honey-Roses, J.; Anguelovski, I.; Bohigas, J.; Chireh, V.; Daher, C.; Konijnendijk, C.; Litt, J.; Mawani, V.; McCall, M.; Orellana, A. The impact of COVID-19 on public space: A review of the emerging questions. *OSF Preprints* **2020**. [CrossRef]
45. Tzoulas, K.; James, P. Peoples' use of, and concerns about, green space networks: A case study of Birchwood, Warrington New Town, UK. *Urban For. Urban Green.* **2010**, *9*, 121–128. [CrossRef]
46. Arora, N.K.; Mishra, J. COVID-19 and importance of environmental sustainability. *Environ. Sustain.* **2020**, *3*, 117–119. [CrossRef]

47. Pfefferbaum, B.; North, C.S. Mental health and the Covid-19 pandemic. *N. Engl. J. Med.* **2020**, *383*, 510–512. [CrossRef] [PubMed]
48. Gualano, M.R.; Lo Moro, G.; Voglino, G.; Bert, F.; Siliquini, R. Effects of COVID-19 lockdown on mental health and sleep disturbances in Italy. *Int. J. Environ. Res. Public Health* **2020**, *17*, 4779. [CrossRef] [PubMed]
49. Park, J.H.; Lee, D.K.; Park, C.; Kim, H.G.; Jung, T.Y.; Kim, S. Park accessibility impacts housing prices in Seoul. *Sustainability* **2017**, *9*, 185. [CrossRef]
50. Le Texier, M.; Schiel, K.; Caruso, G. The provision of urban green space and its accessibility: Spatial data effects in Brussels. *PLoS ONE* **2018**, *13*, e204684. [CrossRef]
51. Nutsford, D.; Pearson, A.L.; Kingham, S. An ecological study investigating the association between access to urban green space and mental health. *Public Health* **2013**, *127*, 1005–1011. [CrossRef]
52. James, P.; Hart, J.E.; Banay, R.F.; Laden, F. Exposure to greenness and mortality in a nationwide prospective cohort study of women. *Environ. Health Perspect.* **2016**, *124*, 1344–1352. [CrossRef]
53. Tian, H.; Liu, Y.; Li, Y.; Wu, C.H.; Chen, B.; Kraemer, M.U.G.; Li, B.; Cai, J.; Xu, B.; Yang, Q.; et al. An investigation of transmission control measures during the first 50 days of the COVID-19 epidemic in China. *Science* **2020**, *368*, 638–642. [CrossRef]
54. Maryland Department of Natural Resources. Available online: https://dnr.maryland.gov/publiclands/Pages/COVID-19-Virus-Alert.aspx (accessed on 15 April 2020).
55. Danley-Greiner, K. MD Beaches Close, Some State Parks Shut Down Due To Coronavirus. *Patch* **2020**. Available online: https://patch.com/maryland/annapolis/md-beaches-close-some-state-parks-shut-down-due-coronavirus (accessed on 30 March 2020).
56. UPDATE: California State Parks Closes Additional Beaches and Parks: Public Reminded to Practice Safe, Physical Distancing. Available online: https://www.parks.ca.gov/NewsRelease/948 (accessed on 20 October 2020).
57. Hall, G.; Laddu, D.R.; Phillips, S.A.; Lavie, C.J.; Arena, R. A tale of two pandemics: How will COVID-19 and global trends in physical inactivity and sedentary behavior affect one another? *Prog. Cardiovasc. Dis.* **2020**. [CrossRef]
58. Rice, W.; Pan, B. Understanding drivers of change in park visitation during the COVID-19 pandemic: A spatial application of Big data. *SocArXiv* **2020**. [CrossRef]
59. Deville, P.; Linard, C.; Martin, S.; Gilbert, M.; Stevens, F.R.; Gaughan, A.E.; Blondel, V.D.; Tatem, A.J. Dynamic population mapping using mobile phone data. *Proc. Natl. Acad. Sci. USA* **2014**, *111*, 15888–15893. [CrossRef] [PubMed]

Publisher's Note: MDPI stays neutral with regard to jurisdictional claims in published maps and institutional affiliations.

Article

Access and Use of Green Areas during the COVID-19 Pandemic: Green Infrastructure Management in the "New Normal"

Yuta Uchiyama and **Ryo Kohsaka** *

Graduate School of Environmental Studies, Nagoya University, Nagoya 464-8601, Japan; uchiyama.yuta@k.mbox.nagoya-u.ac.jp
* Correspondence: kohsaka.ryo@f.mbox.nagoya-u.ac.jp or kohsaka@hotmail.com

Received: 12 October 2020; Accepted: 21 November 2020; Published: 25 November 2020

Abstract: This study aims to identify the influence of the socioeconomic attributes and environmental contexts of citizens' residential areas on the access and use of green areas during the COVID-19 pandemic. The results can aid policymaking and facilitate the safe and unrestricted use of green areas during the pandemic. The access and use of green areas were analyzed using a survey conducted after the official COVID-19 emergency in Japan (16 April to 14 May, 2020). Visits to green areas during the pandemic have gained salience globally from multiple perspectives: health, planning, social justice, and equity. The results of this study demonstrated that socioeconomic factors influenced the frequency of visiting green areas. The factors further influenced the use of the three categories of green areas (parks, agricultural lands, and gardens). Environmental contexts, including the land use patterns in residential areas, also influenced the use of specific types of green areas. Thus, policies need to further facilitate visits to green areas by reflecting the socioeconomic attributes of residents and their households, including income, number of children, gender, and age, incorporating those who have less access and considering the spread of COVID-19 locally. Furthermore, policies for the use of specific green areas, including parks, agricultural lands, and gardens, need to take cognizance of the residents' environmental contexts. Management of specific green areas, like agricultural lands, is required, and residents should be provided with opportunities to use these areas with measures to avoid infection.

Keywords: green area; accessibility; COVID-19; green infrastructure; Japan

1. Introduction

Green areas, including parks, gardens, agricultural lands, and forest lands, provide various ecosystem services—provisional, regulatory, and cultural [1–3]. Agricultural areas and forest lands, including urban spaces, provide us with services, such as food and water. As for the regulatory services, they regulate water and thus help in controlling floods in urban and rural areas. Green areas, including parks, serve residents culturally, as they inspire visitors and provide recreational opportunities. As for the betterment of the ecosystem, appropriate management of green areas and their networks can contribute to urban and rural biodiversity conservation if local governments seek national and international collaborations [4–6]. Despite these fundamental characteristics, citizens who access such services tend to be relatively limited to only certain socioeconomic statuses and environmental contexts [7–10]. Those on a higher income level easily access green areas, while those on low income levels face difficulties due to lack of transportation (including car ownership), time constraints, and entrance or membership fees for such areas [11]. Thus, the socioeconomic status can have an impact on accessibility for certain low-income citizens, as can education level. Furthermore, the environmental context also influences accessibility and use of green areas. For example, properly allocated urban agricultural lands can facilitate agricultural activities [12]. In most areas, even if locals wish to access green areas, it is difficult to do so without proper environmental management and green infrastructure [13]. As fundamental environmental elements that can provide ecosystem services, management of green areas is necessary to enhance the quality of life, for environmental conservation, and to reduce disaster risk in urban and rural areas [14–18].

Basic social attributes such as age, gender, and number of children in households also influence the accessibility of green areas [19,20]. These attributes reflect the regional culture and structure of families and society at large. For instance, in a society with an active aging population [21], the ratio of elderly citizens who use green areas tends to be higher compared with other profiles. As another example, the ratio of females who use green areas might correlate with the number of children in a household in a society with a relatively high ratio of housewives. These societal and cultural contexts need to be considered when developing and implementing green area management. As an infrastructural element of society, green areas need to be supported by the residents. To enhance their ownership and facilitate the citizens' participation in the management, understanding these contexts is essential to establish appropriate policies and actions.

In the ongoing COVID-19 pandemic, trends of access and use of green areas have changed globally [22–24]. National governments have requested that citizens adopt the so-called "new normal" lifestyle to facilitate the changes, including measures such as home offices to avoid crowds in traffic and workplaces. Owing to the lockdown of municipalities to control the spread of infection, access and use of green areas have been restricted. Residents tend to avoid public spaces, and social distancing is required in green areas, especially public parks. The severity of lockdowns differs in various contexts, and Japan was relatively relaxed compared to Europe, China, or India in terms of the strictness of countermeasures and by requesting voluntary cooperation. However, precautionary measures were taken in Japanese parks, including playgrounds, by limiting the facilities. There were social disputes and confusion regarding the extent and perceptions of lockdowns, and some residents made police reports that parks were crowded (the municipalities later asked citizens not to report such incidents, as they were neither urgent nor relevant police tasks).

However, the situations during the COVID-19 pandemic did not necessarily result in negative impacts on residents. For instance, citizens can visit green areas to reduce their COVID-19-related stress issues [25,26]. The circumstances created by COVID-19 can propel the understanding, reexploring, and rediscovering of the meaning and value of green areas. Although green area management in the "new normal" is still in the developing phase, residents' awareness can be enhanced by visiting such areas during this crisis period.

The existing studies tend to analyze the overall number of visitors, but have less data for the visitors' attributes. There need to be more detailed analyses as to "who are the actual visitors?". Given the negative and positive impacts of COVID-19 on green areas, management of such spaces in the "new normal" needs to reflect the socioeconomic attributes and environmental contexts of the residents using them. The influences of these attributes and contexts on the access of green areas during and before the pandemic remain rather unexplored. The purpose of this study is to identify the influence of the socioeconomic attributes of citizens and environmental contexts of their residential areas on the access and use of green spaces. The results can serve as evidence for policymaking to facilitate the safe and equal use of green areas during the pandemic. As for the survey and analysis, the status of access and use of green areas during the time of the COVID-19 crisis is examined using the results of a questionnaire survey. The following sections describe the analysis of the relationships between the access and use of green areas with the socioeconomic attributes of residents and their environmental contexts.

2. Materials and Methods

The results of an online questionnaire survey were used. The survey was conducted after the emergency period (16 April–14 May 2020) of COVID-19 in Japan. During this period, residents were discouraged from going out of their houses and visiting places beyond prefectural borders, although the restrictions in the emergency declaration were not as stringent as lockdowns in other countries. The target site of the survey was Aichi Prefecture and its capital, Nagoya City, which includes one of the largest metropolitan areas and can be a representative case of an urban and rural interface connected with green area networks. The total number of respondents was 1244, with 47.6% females and 52.4% males; the ratio of elderly respondents over 60 years was 36.6%, and other five-year age groups had relatively similar ratios (7–11%), except the age group of 20–24 years (2.9%), which was the youngest age group. The online survey period was from 31 July–1 August 2020.

In the questionnaire, respondents were asked for the following information:

- Socioeconomic attributes: Gender, age, annual household income, number of children in the household;
- Environmental contexts: Zip code area (to compute the ratios of land use categories in individual zip code areas), and whether respondents resided in Nagoya City (Answer: Yes/No);
- Status of access and use of green areas: Whether respondents visited green areas (parks, agricultural lands, common or private gardens, and other green areas) during the emergency period (Answer: Yes/No);
- Change in frequency of access and use of green areas: Whether the frequency of visits was higher compared with the same period in the previous year (Answer: Yes/No).

To compute the ratios of land use categories in zip code areas to analyze the environmental contexts of respondents, Japan Aerospace Exploration Agency (JAXA) High-Resolution Land Use Data (2014–2016) (https://www.eorc.jaxa.jp/ALOS/en/lulc/lulc_index.htm) were used. The resolution of data was a 30 m square grid, and it had 10 land use categories. Because the zip code areas were relatively small, especially in city centers, high-resolution data were used (Figure 1). To understand the ratios of land use categories, the ratios of urban areas, agricultural lands, and forest lands were computed. Nagoya City is a very dense urbanized area, and its environmental context is largely different from those of other municipalities. Considering these characteristics, the place of residence (whether in Nagoya City or not) was included in the environmental context data.

In the data analysis, the Chi-square and *t*-test were performed. The former test was applied to nominal data, and the latter test was applied to continuous data. The status and the change in frequency of access and use of green areas were nominal variables, and the Chi-square test was applied to the analysis of data on socioeconomic attributes, except for age and environmental contexts. Average values of age and ratios of individual land use categories were analyzed using a *t*-test.

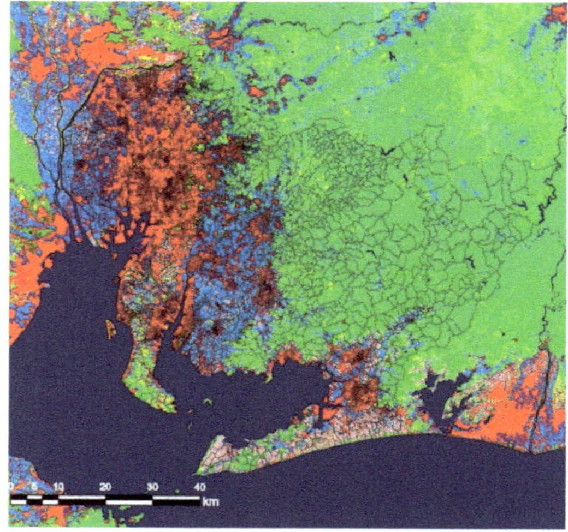

Colour	No.	Class
	1	Water
	2	Urban
	3	Rice paddy
	4	Crop
	5	Grass
	6	Deciduous broadleaf forest
	7	Deciduous needleleaf forest
	8	Evergreen broadleaf forest
	9	Evergreen needleleaf forest
	10	Bare land

Figure 1. Research site (Aichi Prefecture).

3. Results

3.1. Influence of Personal Attributes and Environmental Contexts on Access and Use of Green Areas

This section describes the results of the analysis of relationships between personal attributes and environmental factors and the access and use of green areas. The results of the cross-tabulation of the number of respondents who visited or did not visit green areas and the data of socioeconomic attributes and environmental contexts are found on the left side of Table 1. Statistical analysis was applied to the respondents who reported annual household income (n = 953). The Chi-square and t-test indicate that the number of children in the household and household income variables had statistically significant correlations. As shown in Table 1, the respondents whose households had one or more children tended to visit green areas more compared with the respondents without any children. Furthermore, the respondents whose household income was relatively high (more than 400 million JPY, which is 38 thousand USD (10 November 2020)) tended to visit green areas more compared with those from relatively low-income households.

The t-test was applied to the average values shown on the right side of Table 1, and there was no statistically significant difference between the respondents who did and did not visit green areas during the emergency period.

According to the results, whether respondents visited visit green areas or not was influenced by their socioeconomic attributes, including the number of children and level of household income.

Conversely, environmental context was not a major factor influencing the motivation or action to visit green areas.

3.2. *Influence of Personal Attributes and Environmental Contexts on Use of Individual Types of Green Areas*

This section explains the results of the analysis of the influence of socioeconomic attributes and environmental contexts on the use of individual green areas. The number of respondents to visit parks, agricultural land, and gardens was analyzed, whereas other green areas, including mountains, did not have enough respondents to apply statistical analysis. The Chi-square and *t*-test were applied to the groups of respondents who visited at least one type of green area during the emergency period ($n = 542$).

The results indicated that socioeconomic attributes and environmental contexts that influenced the use of green areas varied among parks, agricultural lands, and gardens (Table 2). Gender, location of residence, age, and ratios of land use categories had a certain influence on park visits. For example, female respondents and residents of Nagoya City tended to visit parks (Table 2). The respondents who visited parks were relatively young compared with those who did not visit. The average values of zip code areas were statistically and significantly varied between the respondents who visited and did not visit parks. Because the zip code areas tended to be larger in rural areas and smaller in urban areas, the results implied that the respondents who visited parks lived in urban areas. This implication was supported by the results of the *t*-test applied for the average values of the ratios of urban areas, agricultural lands, and forest lands. The ratios of those land use categories in zip code areas that are the residential places of respondents who visited parks indicated that the respondents lived in urban areas.

The results revealed that visits to agricultural lands were significantly influenced by location of residence, age, and ratios of land use categories. As for agricultural lands, socioeconomic attributes, except age, were not major factors influencing the motivation and action to visit there. The respondents who visited agricultural lands were relatively old and lived mainly outside Nagoya City (Table 2). The zip code area sizes and the ratios of land use categories implied that they did not live in urban areas. Because of their environmental contexts, those who visited agricultural lands during the emergency period had easy access.

The results indicated that the number of children, location of residence, age, and ratios of land use categories had a certain influence on visiting gardens. As indicated in the Materials and Methods section, "garden" listed in the questionnaire referred to common and private gardens. Respondents who visited gardens tended to have one or more children. Respondents who visited gardens tended to reside outside Nagoya City (Table 2). The ratios of urban areas were relatively lower, and the ratios of agricultural lands were higher in areas of their zip code and residence. These results implied that the respondents who visited gardens tended to live in non-urbanized areas. The average age of the respondents who visited gardens was slightly older than those who did not visit.

As an overall trend, parks were visited by relatively younger respondents who lived in urban areas, and agricultural lands and gardens were visited by the older respondents who lived in rural areas or areas in between. Income level was not a major factor influencing motivation and action to visit the three individual types of green areas.

3.3. *Change of Frequency of Using Green Areas*

The results of the analysis of relationships between personal attributes and environmental contexts and change in frequency of visiting green areas indicated a similar trend to that in Section 3.1, which revealed the analysis results of these factors' influence on access and the use of green areas. Environmental context was not a major factor that influenced the change in frequency of visiting green areas. Conversely, socioeconomic attributes were related to the change in the frequency (Table 3). Household income was the common major factor that influenced the analysis results provided in Section 3.1. The respondents' gender was a factor that influenced the results of this section, and while the number of children was not a factor, it did influence the respondents visiting green areas, as shown in Section 3.1.

Table 1. Relationships between personal attributes and environmental factors and access and use of green areas.

	Number of Respondents								Average Value				
	Number of Children (>1)	Number of Children (0)	Household Income (≥400 Million JPY)	Household Income (<400 Million JPY)	Male	Female	Nagoya City	Other Municipality	Age	Zip Code Area (ha)	Urban Area (%)	Agricultural Land (%)	Forest Land (%)
Yes	439	159	421	177	343	255	190	408	51.3	165.6	66.0	23.3	5.0
No	215	140	229	126	208	147	111	244	52.3	149.4	67.2	22.8	4.2

(Green: $p < 0.01$, orange: $p < 0.1$; 400 million JPY is equal to 38 thousand USD (10 November 2020); Yes: accessed and used, No: did not access and use).

Table 2. Relationships between personal attributes and environmental factors and access and use of the three types of green areas.

	Number of Respondents								Average Value				
	Number of Children (>1)	Number of Children (0)	Household Income (≥400 Million JPY)	Household Income (<400 Million JPY)	Male	Female	Nagoya City	Other Municipalities	Age	Zip Code Area (ha)	Urban Area (%)	Agricultural Land (%)	Forest Land (%)
Park (y)	142	51	139	54	103	90	71	122	47.0	121.3	71.7	19.3	3.2
Park (n)	261	88	242	107	213	136	101	248	54.7	184.4	63.2	25.2	6.0
Agricultural Land (y)	20	4	17	7	17	7	2	22	64.1	565.5	46.2	31.7	17.2
Agricultural Land (n)	383	135	364	154	299	219	170	348	51.4	143.2	67.2	22.7	4.5
Garden (y)	180	43	152	71	122	101	52	171	53.3	172.6	60.5	28.2	5.6
Garden (n)	223	96	229	90	194	125	120	199	51.0	154.5	70.3	19.6	4.6

(Green: $p < 0.01$, yellow: $p < 0.05$, orange: $p < 0.1$; 400 million JPY is equal to 38 thousand USD (10 November 2020); (y): accessed and used, (n): did not access and use).

Table 3. Relationships between personal attributes and environmental factors and change in frequency of visiting green areas.

	Number of Respondents								Average Value				
	Number of Children (>1)	Number of Children (0)	Household Income (≥400 Million JPY)	Household Income (<400 Million JPY)	Male	Female	Nagoya City	Other Municipality	Age	Zip Code Area (ha)	Urban Area (%)	Agricultural Land (%)	Forest Land (%)
Frequency (Higher)	81	35	92	24	57	59	35	81	47.2	170.0	66.8	23.9	4.3
Frequency (Stable/Lower)	322	104	289	137	259	167	137	289	53.3	159.7	66.1	22.9	5.2

(Green: $p < 0.01$, yellow: $p < 0.05$; 400 million JPY is equal to 38 thousand USD (10 November 2020)).

In summary, the household income of the respondents whose frequency of visiting green areas tended to be higher than that of the respondents whose visits were stable or lower. Regarding gender, the frequency of female respondents visiting green areas tended to be higher than that of males. The average age of the respondents who frequently visited green areas was relatively higher (Table 3).

Household income was detected as a factor that influenced (i) the change in frequency of visiting green areas (Section 3.3) and (ii) whether the respondents visited green areas or not (Section 3.1). Furthermore, it should be noted that environmental contexts that influenced the use of the three types of green areas were not major factors regarding (i) the change in frequency and (ii) visiting green areas.

In the next section, the interpretation of the results provided in Section 3 and implications for policies and management of green areas will be discussed.

4. Discussion and Conclusions

This study describes a major behavioral pattern seen during the COVID-19 crisis in Japan. We found that socioeconomic attributes influenced the changes in frequency of visiting green areas. The attributes further influenced whether the respondents did or did not visit green areas, categorized into three types. Alternatively, environmental contexts, including the residential location of the respondents, was a factor that had a certain influence only on the use of the three types of green areas. These results implied that motivation or actions of visiting green areas tended to be influenced by the socioeconomic attributes of the residents. Environmental contexts might have an influence on the motivation of visiting green areas by providing scenery and an opportunity to visit green areas near the citizens' residential spaces; however, the influence might not be relatively strong compared with that of socioeconomic attributes. The use of specific green areas such as parks, agricultural lands, and gardens tended to be influenced by socioeconomic attributes and environmental contexts. Future policies to promote visits to green areas or to enhance the frequency of visits need to focus on the socioeconomic attributes of residents, such as household income, number of children, gender, and age, rather than environmental ones. These factors influence whether residents will visit green areas or not.

Before the pandemic, a study focusing on a Japanese city suggested that household income and related socioeconomic attributes did not correlate with the use of green areas [27]. Although further research is needed to detect the specific impact of income level on the access and use of green areas, this study showed that income level might have become a factor that influences the use of green areas in Japan. In Europe, socioeconomic attributes, including income level and age, were detected as factors influencing the access of green areas even before the period [28]. As for research in North America, existing studies showed the correlation between high socioeconomic status and access of green areas [8,9]. In China, those attributes were not influential factors on the access of green areas [29], which is a similar trend to that shown in Rupprecht et al. [27]. As for the research published right after the beginning of the pandemic, it was reported that the frequency and access of green areas had changed, as the existing studies, such as Žlender and Thompson [30], analyzed accessibility and purpose of visit [22,24,31]. Some of the research focused on health issues rather than socioeconomic issues [26,32]. As Honey-Rosés et al. [23] suggested, inequities and exclusions can be included in research on access and use of green areas in the pandemic, and this research provided the empirical results that can contribute to these topics. Further empirical research is needed to verify the status and trends of the inequities and exclusions in the access and use of green areas in different regions and countries with different backgrounds, such as post-socialist countries [10].

The question remains as to how one promotes visits from those social stratifications that are not actively visiting the areas. Reflecting on the opinions from the citizens with different socioeconomic attributes, policies and actions are necessary to enhance the accessibility of green areas for residents whose socioeconomic status is relatively low. As for the gender and number of children, they seem not to correlate in the results, although both are factors that influence access and use of green areas. It can be assumed that certain numbers of female respondents visited green areas with their children, but that trend was not very clear in the results. In future research, the detailed influence of gender and number of children on access to green areas needs to be identified, and specific policies that can be applied for children, parents, and other family members need to be discussed. The involvement of citizens with different backgrounds is needed in the process of policymaking and practices of green area management [7].

Having said this, the environmental contexts of residents are influential once the residents decide to visit the green areas. Therefore, policies regarding the use of specific green areas, including parks, agricultural lands, and gardens, need to be considered with the general policies facilitating visits to green areas. The results suggest that parks were mainly used by urban residents, and agricultural lands and gardens were used by rural residents. If there is a policy demand for facilitating the use of specific green areas, such as agricultural lands, both adequately preparing the green areas and providing opportunities for the use of such areas are required.

Author Contributions: Writing—original draft, Y.U.; Writing—review and editing, R.K. All authors have read and agreed to the published version of the manuscript.

Funding: This research was funded by the Japan Society for the Promotion of Science: JSPS KAKENHI Grant Numbers: JP16KK0053; JP17K02105; JP20K12398, as well as the JST-RISTEX Project "Development and Implementation of Consensus Building Method for Policies on Balanced Conservation, Agriculture and Forestry", the Kurita Water and Environment Foundation: 20C002, and the Daiko Foundation: 2019.

Conflicts of Interest: The authors declare no conflict of interest.

References

1. Bolund, P.; Hunhammar, S. Ecosystem services in urban areas. *Ecol. Econ.* **1999**, *29*, 293–301. [CrossRef]
2. Kohsaka, R. Developing biodiversity indicators for cities: Applying the DPSIR model to Nagoya and integrating social and ecological aspects. *Ecol. Res.* **2010**, *25*, 925–936. [CrossRef]
3. Larondelle, N.; Haase, D. Urban ecosystem services assessment along a rural–urban gradient: A cross-analysis of European cities. *Ecol. Indic.* **2013**, *29*, 179–190. [CrossRef]
4. Bai, X. Integrating global environmental concerns into urban management: The scale and readiness arguments. *J. Ind. Ecol.* **2007**, *11*, 15–29. [CrossRef]
5. Wilkinson, C.; Sendstad, M.; Parnell, S.; Schewenius, M. Urban governance of biodiversity and ecosystem services. In *Urbanization, Biodiversity and Ecosystem Services: Challenges and Opportunities*; Springer: Dordrecht, The Netherland, 2013; pp. 539–587.
6. Uchiyama, Y.; Hayashi, K.; Kohsaka, R. Typology of cities based on city biodiversity index: Exploring biodiversity potentials and possible collaborations among Japanese cities. *Sustainability* **2015**, *7*, 14371–14384. [CrossRef]
7. Kabisch, N.; Haase, D. Green justice or just green? Provision of urban green spaces in Berlin, Germany. *Landsc. Urban Plan.* **2014**, *122*, 129–139. [CrossRef]
8. Rigolon, A. A complex landscape of inequity in access to urban parks: A literature review. *Landsc. Urban Plan.* **2016**, *153*, 160–169. [CrossRef]
9. Rigolon, A.; Browning, M.; Jennings, V. Inequities in the quality of urban park systems: An environmental justice investigation of cities in the United States. *Landsc. Urban Plan.* **2018**, *178*, 156–169. [CrossRef]
10. Kronenberg, J.; Haase, A.; Łaszkiewicz, E.; Antal, A.; Baravikova, A.; Biernacka, M.; Dushkova, D.; Filčak, R.; Haase, D.; Ignatieva, M.; et al. Environmental justice in the context of urban green space availability, accessibility, and attractiveness in postsocialist cities. *Cities* **2020**, *106*, 102862. [CrossRef]
11. Marquet, O.; Hipp, J.A.; Alberico, C.; Huang, J.H.; Fry, D.; Mazak, E.; Lovasi, G.S.; Floyd, M.F. Park use preferences and physical activity among ethnic minority children in low-income neighborhoods in New York City. *Urban For. Urban Green.* **2019**, *38*, 346–353. [CrossRef]
12. Rich, K.M.; Rich, M.; Dizyee, K. Participatory systems approaches for urban and peri-urban agriculture planning: The role of system dynamics and spatial group model building. *Agric. Syst.* **2018**, *160*, 110–123. [CrossRef]
13. Venkataramanan, V.; Packman, A.I.; Peters, D.R.; Lopez, D.; McCuskey, D.J.; McDonald, R.I.; Miller, W.M.; Young, S.L. A systematic review of the human health and social well-being outcomes of green infrastructure for stormwater and flood management. *J. Environ. Manag.* **2019**, *246*, 868–880. [CrossRef] [PubMed]

14. Elbakidze, M.; Angelstam, P.; Yamelynets, T.; Dawson, L.; Gebrehiwot, M.; Stryamets, N.; Johansson, K.E.; Garrido, P.; Naumov, V.; Manton, M. A bottom-up approach to map land covers as potential green infrastructure hubs for human well-being in rural settings: A case study from Sweden. *Landsc. Urban Plan.* **2017**, *168*, 72–83. [CrossRef]
15. Lin, B.B.; Philpott, S.M.; Jha, S.; Liere, H. Urban agriculture as a productive green infrastructure for environmental and social well-being. In *Greening Cities*; Springer: Singapore, 2017; pp. 155–179.
16. Kohsaka, R.; Uchiyama, Y. Motivation, strategy and challenges of conserving urban biodiversity in local contexts: Cases of 12 municipalities in Ishikawa, Japan. *Procedia Eng.* **2017**, *198*, 212–218. [CrossRef]
17. Uchiyama, Y.; Kohsaka, R. Application of the City Biodiversity Index to populated cities in Japan: Influence of the social and ecological characteristics on indicator-based management. *Ecol. Indic.* **2019**, *106*, 105420. [CrossRef]
18. Yoo, S.Y.; Kim, T.; Ham, S.; Choi, S.; Park, C.R. Importance of Urban Green at Reduction of Particulate Matters in Sihwa Industrial Complex, Korea. *Sustainability* **2020**, *12*, 7647. [CrossRef]
19. Wood, L.; Hooper, P.; Foster, S.; Bull, F. Public green spaces and positive mental health–investigating the relationship between access, quantity and types of parks and mental wellbeing. *Health Place* **2017**, *48*, 63–71. [CrossRef]
20. Sikorska, D.; Łaszkiewicz, E.; Krauze, K.; Sikorski, P. The role of informal green spaces in reducing inequalities in urban green space availability to children and seniors. *Environ. Sci. Policy* **2020**, *108*, 144–154. [CrossRef]
21. Yung, E.H.; Conejos, S.; Chan, E.H. Social needs of the elderly and active aging in public open spaces in urban renewal. *Cities* **2016**, *52*, 114–122. [CrossRef]
22. Venter, Z.; Barton, D.; Gundersen, V.; Figari, H.; Nowell, M. Urban nature in a time of crisis: Recreational use of green space increases during the COVID-19 outbreak in Oslo, Norway. *Environ. Res. Lett.* 2020. Available online: https://iopscience.iop.org/article/10.1088/1748-9326/abb396/pdf (accessed on 23 November 2020).
23. Honey-Rosés, J.; Anguelovski, I.; Bohigas, J.; Chireh, V.; Daher, C.; Konijnendijk, C.; Litt, J.; Mawani, V.; McCall, M.; Orellana, A.; et al. The impact of COVID-19 on public space: An early review of the emerging questions–design, perceptions and inequities. *Cities Health* **2020**, 1–17. [CrossRef]
24. Derks, J.; Giessen, L.; Winkel, G. COVID-19-induced visitor boom reveals the importance of forests as critical infrastructure. *For. Policy Econ.* **2020**, *118*, 102253. [CrossRef] [PubMed]
25. Freeman, S.; Eykelbosh, A. COVID-19 and outdoor safety: Considerations for use of outdoor recreational spaces. *Natl. Collab. Centre Environ. Health.* 2020. Available online: https://ncceh.ca/sites/default/files/COVID-19%20Outdoor%20Safety%20-%20April%2016%202020.pdf (accessed on 16 April 2020).
26. Xie, J.; Luo, S.; Furuya, K.; Sun, D. Urban Parks as Green Buffers During the COVID-19 Pandemic. *Sustainability* **2020**, *12*, 6751. [CrossRef]
27. Rupprecht, C.D.; Byrne, J.A.; Ueda, H.; Lo, A.Y. 'It's real, not fake like a park': Residents' perception and use of informal urban green-space in Brisbane, Australia and Sapporo, Japan. *Landsc. Urban Plan.* **2015**, *143*, 205–218. [CrossRef]
28. Wüstemann, H.; Kalisch, D.; Kolbe, J. Access to urban green space and environmental inequalities in Germany. *Landsc. Urban Plan.* **2017**, *164*, 124–131. [CrossRef]
29. Wei, F. Greener urbanization? Changing accessibility to parks in China. *Landsc. Urban Plan.* **2017**, *157*, 542–552. [CrossRef]
30. Žlender, V.; Thompson, C.W. Accessibility and use of peri-urban green space for inner-city dwellers: A comparative study. *Landsc. Urban Plan.* **2017**, *165*, 193–205. [CrossRef]
31. Ugolini, F.; Massetti, L.; Calaza-Martínez, P.; Cariñanos, P.; Dobbs, C.; Ostoic, S.K.; Marin, A.M.; Pearlmutter, D.; Saaroni, H.; Šaulienė, I.; et al. Effects of the COVID-19 pandemic on the use and perceptions of urban green space: An international exploratory study. *Urban For. Urban Green.* **2020**, *56*, 126888. [CrossRef]
32. Shoari, N.; Ezzati, M.; Baumgartner, J.; Malacarne, D.; Fecht, D. Accessibility and allocation of public parks and gardens in England and Wales: A COVID-19 social distancing perspective. *PLoS ONE* **2020**, *15*, e0241102. [CrossRef]

Publisher's Note: MDPI stays neutral with regard to jurisdictional claims in published maps and institutional affiliations.

Article

How Blue Carbon Ecosystems Are Perceived by Local Communities in the Coral Triangle: Comparative and Empirical Examinations in the Philippines and Indonesia

Jay Mar D. Quevedo [1], Yuta Uchiyama [2], Kevin Muhamad Lukman [1] and Ryo Kohsaka [2,*]

[1] Graduate School of Environmental Studies, Tohoku University, Sendai City 980-8572, Japan; quevedojaymar@gmail.com (J.M.D.Q.); kevin.muhamad.lukman@gmail.com (K.M.L.)

[2] Graduate School of Environmental Studies, Nagoya University, Nagoya City 464-8601, Japan; uchiyama.yuta@k.mbox.nagoya-u.ac.jp

* Correspondence: kohsaka@hotmail.com

Abstract: Blue carbon ecosystem (BCE) initiatives in the Coral Triangle Region (CTR) are increasing due to their amplified recognition in mitigating global climate change. Although transdisciplinary approaches in the "blue carbon" discourse and collaborative actions are gaining momentum in the international and national arenas, more work is still needed at the local level. The study pursues how BCE initiatives permeate through the local communities in the Philippines and Indonesia, as part of CTR. Using perception surveys, the coastal residents from Busuanga, Philippines, and Karimunjawa, Indonesia were interviewed on their awareness, utilization, perceived threats, and management strategies for BCEs. Potential factors affecting residents' perceptions were explored using multivariate regression and correlation analyses. Also, a comparative analysis was done to determine distinctions and commonalities in perceptions as influenced by site-specific scenarios. Results show that, despite respondents presenting relatively high awareness of BCE services, levels of utilization are low with 42.9–92.9% and 23.4–85.1% respondents in Busuanga and Karimunjawa, respectively, not directly utilizing BCE resources. Regression analysis showed that respondents' occupation significantly influenced their utilization rate and observed opposite correlations in Busuanga (positive) and Karimunjawa (negative). Perceived threats are found to be driven by personal experiences—occurrence of natural disasters in Busuanga whereas discerned anthropogenic activities (i.e., land-use conversion) in Karimunjawa. Meanwhile, recognized management strategies are influenced by the strong presence of relevant agencies like non-government and people's organizations in Busuanga and the local government in Karimunjawa. These results can be translated as useful metrics in contextualizing and/or enhancing BCE management plans specifically in strategizing advocacy campaigns and engagement of local stakeholders across the CTR.

Keywords: blue carbon ecosystems; coral triangle; Philippines; Indonesia; local perceptions

Citation: Quevedo, J.M.D.; Uchiyama, Y.; Muhmad Lukman, K.; Kohsaka, R. How Blue Carbon Ecosystems Are Perceived by Local Communities in the Coral Triangle: Comparative and Empirical Examinations in the Philippines and Indonesia. *Sustainability* **2021**, *13*, 127. https://dx.doi.org/10.3390/su13010127

Received: 19 October 2020
Accepted: 22 December 2020
Published: 24 December 2020

Publisher's Note: MDPI stays neutral with regard to jurisdictional claims in published maps and institutional affiliations.

1. Introduction

Coastal ecosystems are among the most productive ecosystems; offering beneficial services that enhance people's well-being and supporting local communities and national economies [1,2]. Among these services are food provision, habitat for commercially important species [3], coastal protection [4], and cultural services [5]. Another key service coastal wetlands provide that was overlooked in the past is the regulation of the global climate; currently referred to as "blue carbon" [6]. This concept recognizes the vital role of the coastal wetlands as buffers to the adverse effects of the changing, in this case, increasing, world atmospheric carbon dioxide levels [7]. This is possible through the thriving photosynthetic organisms in coastal ecosystems that extract carbon dioxide directly from the atmosphere and surface waters [8,9]. Mangroves, seagrass meadows, and tidal marshes, collectively called blue carbon ecosystems (BCEs), sequester and store carbon dioxide as organic carbon in their biomass (above and below ground) and soil material [10,11]. BCEs may carry out this process continuously for over thousands of years, locking away carbon that could contribute to the heating of the earth's atmosphere into a large number of carbon stocks in biomass and organic-rich soils [6].

One of the world's richest hotspots of marine biodiversity lies in the Coral Triangle Region (CTR) [12], where Indonesia and the Philippines are geographically located. In terms of BCEs in the region, Indonesia has the largest mangrove and seagrass coverage, around 2,707,572 ha [13] and 3,000,000 ha [14], respectively, whereas the Philippines has roughly 256,185 ha of mangrove forests [15] and 97,800 ha of seagrass meadows [16]. Tidal marshes are not found in these countries since they thrive in mid- to high-latitude regions [17]; ranging from arctic to subtropical regions [18]. The degradation of coastal wetlands in the Philippines and Indonesia could result in discernable loss of beneficial services such as food sources, protection against storm surges, and cultural services [19–21]. Moreover, as BCEs are in decline, their intangible and long-term service of carbon sequestration and storage are likely to be affected as well. A large number of carbon stocks sequestered over the years in Indonesia and the Philippines are threatened to be released back into the air, contributing to the further rising of global temperature, if these ecosystems continue to be degraded and lost [22,23]. Much of the deterioration of these coastal resources is mostly anthropogenic, particularly due to land-use conversion [1,24,25].

In response, researchers, policymakers, and implementers are gearing towards building a strong foundation of science, policy, and sustainable coastal management practices for the conservation and restoration of BCEs as a means of collective effort in addressing climate change [6]. Some of the recent advances on blue-carbon-related studies in the Philippines and Indonesia include carbon stock assessment and carbon sequestration potential [14,26,27], policies and finance mechanisms [28–30], and public's perceptions [31–34]. Collaborative workshops and other initiatives on blue carbon involving these two countries along with other members in the CTR are also progressing (i.e., [6,35]). There is the International Partnership for Blue Carbon (IPBC) established during the Paris Agreement, UNFCCC COP21, in 2015, which was aimed at connecting efforts of research organizations, governments, non-government and international organizations in enhancing the protection and restoration of BCEs [36]. Such partnerships can foster and strengthen the relationships of the involved countries by sharing knowledge, experiences, and expertise in understanding better the importance of BCEs in global climate regulation and adaptation, achieving sustainable development goals, growing the blue economy, and meeting national commitments to the Paris Agreement [6].

Despite the gaining momentum of the "blue carbon" discourse and collaborative actions in the international and national arenas, more work is still needed at the local level especially where local governments, implementers, and residents are the ones interacting with—benefiting from and taking advantage of resources from—these ecosystems. This work pursues how BCE initiatives permeate through the local communities in CTR, particularly in the countries of Indonesia and the Philippines using perception inquiries. Local perceptions, based on comprehensive theoretical and empirical evidence, have a critical role in supporting collective responses for the sustainable management of natural resources (i.e., [34,37]). Engaging local communities in ecosystem service assessments helps define their role in the multi-governance of environments as well as the importance of ecosystem services (ES) and the factors that influence social preferences and trade-offs related to land-use change and decision-making [38]. ES in this study is based on the definition of [38] which refers to the benefits people obtain from ecosystems. These include provisioning services (i.e., food), regulating services (i.e., climate regulation), cultural services (i.e., recreational benefits), and supporting services (i.e., habitat). Local's perception of ES is a very subjective process—it can be based on their comprehension, interpretation, and experiences. There is also a concern of how ES provides well-being and how these benefits are valued, whether as an "instrumental" value when attributed to a particular purpose or as a "relational" value when used to measure certain types of interactions, by society [39].

For instance, some locals may recognize cultural services based on their aesthetic, educational, and therapeutic values [40] while others based on the accessibility and proximity of the resources (i.e., tourism sites) [41]. There is also a shift at the conceptual level which is focusing more on "indirect drivers" in ES assessments although "direct drivers" are still largely highlighted in frameworks such as the Driver-Pressure-State-Impact-response (DPSIR) [42,43]. Indirect drivers (i.e., demographic, economic, socio-political, cultural) can heavily influence locals' perceptions and attitudes towards the environment with subsequent environmental implications (positive or negative).

In this study, people's awareness level, utilization rates, perceived threats, and management strategies were gathered and used as proxies in determining the familiarity of coastal communities to BCEs, which in turn could reflect the current management directives at a local scale in the Philippines and Indonesia in CTR. The people's awareness and utilization of ES in this work are closely associated with assessing the instrumental value of the benefits (i.e., food provision, recreation) since these benefits allow people to achieve a good quality of life [39]. "Relational" values, which are equally important, were not assessed since this involved a thorough examination of the relationship between the people and nature, like determining their specific principles or moral duties on how they can relate to nature [39]. Socio-demographic characteristics were factored in, since previous studies have shown their effect on an individual's perceptions [33,37]. Furthermore, a comparative analysis was done to determine distinctions and commonalities in perceptions as influenced by site-specific scenarios. These perceptions can be translated to become useful metrics in contextualizing and/or enhancing coastal management plans specifically in strategizing advocacy campaigns and engagement of local stakeholders. Up until recently, there has been a preference for a bottom-up approach in management and governance in the Philippines and Indonesia. How the study underscores local perceptions could also contribute to this grassroots/community-based and informed course of action for the national level to highly consider and prioritize. The availability of this information across the CTR could be a sound foundation to compare and contrast how coastal communities from different countries perceive and value their resources. This crucial learning could then be furthered into identifying common grounds that can be transferrable across or translated into a contextualized regional program within the CTR.

2. Research Methods

2.1. Study Sites

This study was conducted in the municipality of Busuanga in Busuanga Island, Palawan province, Philippines, and Karimunjawa Island, Jepara Regency, Indonesia (Figure 1). The two sites present a good opportunity to show how locals perceive BCEs, their services, and management status because of large communities depending on them. Both sites have relatively the same characteristics (i.e., economic activities, presence of a zonation system, and different stakeholders) that could potentially influence communities' perceptions. These site-specific settings can reflect how BCEs are locally managed, which is critical to understanding and achieving sustainability goals across the CTR.

In terms of economic activities, 70% of the activities in Busuanga town comes from fishing, forestry, and agriculture [44] while 74% comes from fishing and farming in Karimunjawa Island [45]. Fishing-related activities in the sites can be supported by the presence of BCEs: the former has an estimated 4738 ha of mangroves and 3726 ha of dense and sparse seagrass beds [44] while the latter has roughly 400 ha of mangroves and 404 ha of seagrass meadows [46]. Despite the economic importance, BCEs in the sites are still subjected to a lot of human-derived stresses (i.e., illegal cutting), thus, several management schemes are in place to protect and conserve them. For instance, Busuanga Island has been identified as a partially protected key biodiversity area (KBA) along with Calauit, Culion, and Coron Islands [47] with an ecological zoning plan being followed under the strategic environmental plan (SEP or Republic Act No. 7611). The environmental conservation in the island is, among others, governed by a special institution called the Palawan Council for Sustainable Development (PCSD). Meanwhile, Karimunjawa Island, along with the 26 islands, comprise the Karimunjawa National Park (KNP), a protected marine park, which was established under the Plantation Decree no. 78/kpts-II/1999 by the Ministry of Forestry [48]. A stringent zoning system that consists of eight zones is implemented on the island [48]. The zonation system in the sites governs where and how the coastal resources should be used.

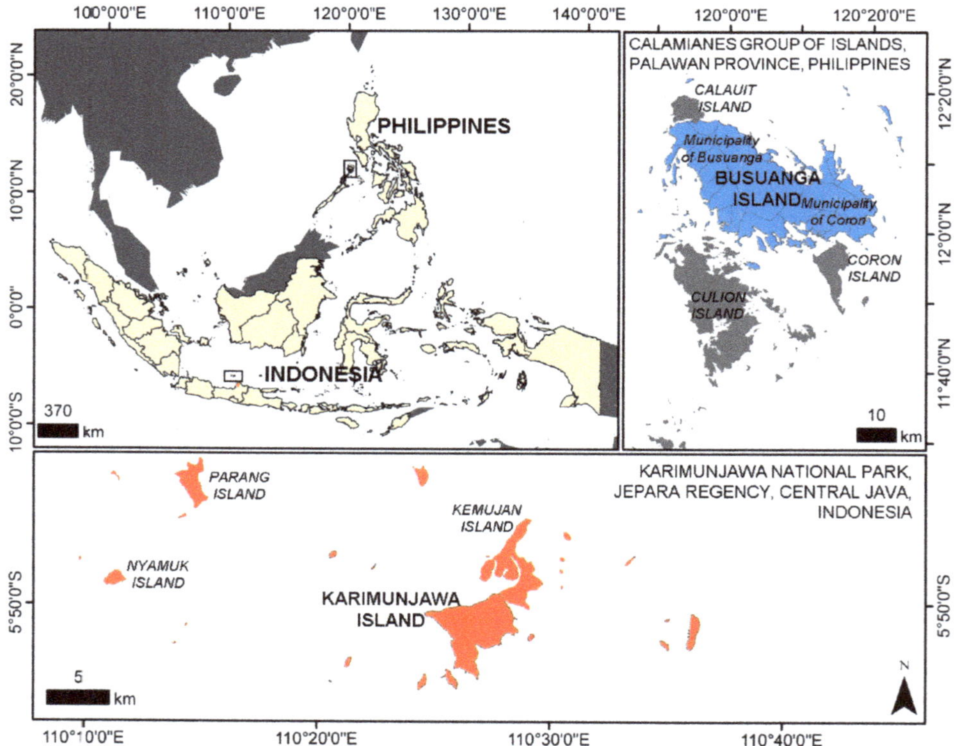

Figure 1. Location map of the study sites.

The sites also reflect the potential role of different stakeholders like local residents, people's organizations (POs), non-government organizations (NGOs), government agencies, and private sectors in the management of BCEs. In Busuanga, the municipal agriculture office (MAO) is in charge of the management of land and marine resources of the town. Following the mandates from the provincial and national offices, the MAO oversees programs and collaborations with other agencies. The MAO directly supervises the barangay fisheries and aquatic resources management council (BFARMC) and coordinates with POs (i.e., fishers' associations). The presence of these grassroots-led groups encourages the active participation of the community in management-related activities [49]. Another important stakeholder present in Busuanga is C3 Philippines, an NGO which empowers local communities to manage sustainably their coastal resources. These various groups in Busuanga are important in achieving a better integrated coastal management system [49]. Meanwhile, in Karimunjawa, the work of [50] has documented that 67.74% of management-related activities in the island are performed by relevant government agencies like Balai Taman Nasional Karimunjawa while a small fraction belongs to residents or community and NGOs, about 22.6% and 8.6% of the activities, respectively. Government agencies usually lead and collaborate with other organizations to conduct activities on the island.

For instance, Karimunjawa National Park Authority has collaborated with NGOs like the Wildlife Conservation Society (WCS), Taka (local), and the University Diponegoro in conducting spatial planning and stakeholder consultation processes to revise the zoning system [48]. On the other hand, community-led and NGO-led activities are often limited since they are dependent on funding agencies [50].

2.2. Sampling and Survey Procedures

Identification of the survey sites followed the criteria of [33]: the presence of BCEs, the proximity of coastal communities, and accessibility of the village. The interviews were conducted in the municipality of Busuanga in Busuanga Island and in Karimunjawa Island in KNP (Figure 1). Given the set conditions, the respondents were selected randomly; surveying one household in every five-household interval where possible [33,34]. Also, stating the purpose of the survey and asking the permission of each respondent was carefully observed. Furthermore, field enumerators who are associated with the respective government units assisted in the conduct of the surveys. For clarity and consistency purposes in presenting the results and discussions, the sites will be delineated simply as "Busuanga" and "Karimunjawa".

The sample size computation was carefully done to get an appropriate representation of the populations and reliable inferences [51]. Although a sample size with the smallest margin of error (i.e., 2–5%) will always be considered a good representative of the population [52], the level of precision could also depend on the amount of risk a researcher is willing to accept [51]. Thus, in this study, we calculated the sample size in each site at a 95% confidence level with a 10% sampling error using Cochran's formula [53]: $n = n_0/(1 + n_0/N)$, where $n_0 = (t^2 * p * q)/d^2$, and t = value of selected alpha level (in this study, α is 0.05, so the critical value is 1.96), p = estimated proportion of the population which has the attribute in question, $q = 1 - p$, and d = acceptable margin of error (in this study, 0.10), and N = population size (22,046 for Busuanga [54] and 9249 for Karimunjawa [45]). The confidence interval of 95% and the margin of error of 10% were selected based on previous works on household surveys (i.e., [33,34]).

A total of 98 locals were randomly interviewed in Busuanga from 19–25 July 2019 whereas only 47 residents were surveyed in Karimunjawa from 8–12 October 2019. Due to limited time in the field survey in Karimunjawa, we were only able to interview 47 respondents instead of 95 people, which is based on our sample size calculation. Due to this, the sampling error for Karimunjawa has increased to 14% from the planned 10% margin of error. We accepted this risk and limitations in accordance with the principles of [51] and the value of internationally comparative survey results was considered and the results were used in this study, carefully considering the difference in the levels of precision.

2.3. Survey Questionnaire

The questionnaire used was translated to the local language of the study sites, Filipino/Tagalog for Busuanga and Bahasa for Karimunjawa. It has four sections, namely (A) socio-demographics, (B) awareness level, and (C) utilization frequency of ES, and (D) perceived threats and management strategies (modified from [33,34]). Section A profiled the name, gender, age, residency, education, and occupation of the respondents. Section B and C used a five-point Likert scale; awareness level was measured from 1 (not aware) to 5 (extremely aware) while utilization rate is weighed from 1 (never use) to 5 (every day). Section D measures the perceived threats and management priorities through a rating with most (1) to least (10) damaging threat and top (1) to least (7) priority, respectively. Lastly, the respondents were asked to identify from a list of stakeholders who should be managing the BCEs in their areas.

2.4. Data Analysis

This study utilized descriptive statistics (i.e., frequencies, percentages, and means), comparison tests, multivariate regressions, and correlations. Descriptive statistics were used to show the respondents' socio-demographic characteristics, awareness, and utilization of BCE services, and their perceptions on threats and management of these resources. The Mann–Whitney U-test was used to evaluate if there are significant differences between the study sites in locals' awareness level and utilization frequency, and perceptions of threats and management strategies of BCEs. Moreover, multivariate regressions were utilized to evaluate the influence of respondents' socio-demographic characteristics on their awareness and utilization of BCE services. Lastly, Spearman's rank correlation coefficient (ρ) was carried out to evaluate whether the locals' utilization patterns can be linked with their knowledge of ES [33].

3. Results

3.1. Socio-Demographic Profile of the Respondents

Table 1 shows the socio-demographics of the respondents from Busuanga (n = 98) and Karimunjawa (n = 47). The respondents in Busuanga are almost equally distributed in terms of gender (male is 51.0% and female is 49.0%) with a mean age of 44 years old. The majority (75.5%) of the respondents is living in the neighborhood since birth (21 years or more) while others, about 11.2% are relatively new in the area (5–10 years). In terms of formal education, 51.0% of the locals have finished primary school while 30.6% have completed secondary school. About 11.2% of the respondents did not finish formal education. Moreover, in terms of occupation, salaried individuals (daily, weekly, or monthly earners) covering part-time workers, skilled workers, and government employees comprised 40.8% of the total interviewees while 23.5% are fishermen and 5.1% are farmers. The unemployed group accounts for 30.6% of the total respondents.

In contrast, a majority (87.2%) of the respondents in Karimunjawa Island are male while only 12.8% are female respondents. More than half (51.1%) of the respondents are in the "41 to 50 years old" age group, with an average age of 40 years old. About 57.4% of the respondents are living on the island since birth while a few (4.3% to 6.4%) have just resided in the area for less than 10 years. In terms of education, 66.0% of the respondents have completed secondary school, 23.4% finished primary education and about 10.6% are degree holders. The occupation of the respondents is distributed to salaried individuals (34.0%) who include merchants, government employees, and skilled workers, fishermen (21.3%), and housewives (6.4%). The remaining 38.3% did not divulge their occupation information.

3.2. Respondents' Awareness of Blue Carbon Ecosystem Services

The awareness level of mangrove ES in Busuanga is fairly consistent where 27.6% to 35.7% of the respondents in Busuanga are "very aware" of all the ES (i.e., source of food, coastal protection, carbon sequestration, a habitat of many organisms, cultural services) listed in the questionnaire while only 10.2% to 22.4% are not aware of these benefits (Figure 2). Meanwhile, the recognition of seagrass ES depends on the type of service. Seagrass beds as a source of food, habitat, nursery, feeding and breeding ground of many organisms, and site for cultural activities are highly recognized ("moderate" to "extremely aware") by 50.0% to 60.2% of the respondents while regulating services like coastal protection and natural buffer are poorly known ("not aware") by 42.9% to 45.9% of the sample size. Another regulating service that the locals are not so familiar with is the capacity of seagrasses to sequester and store carbon; a little over half (54.1%) of them are aware while 45.9% are "slightly aware" to "not aware".

Table 1. Socio-demographic profile of the respondents from Busuanga and Karimunjawa.

Socio-Demographic Profile	Busuanga (n = 98)		Karimunjawa (n = 47)	
	Frequency	Percentage	Frequency	Percentage
Gender				
Male	50	51.0	41	87.2
Female	48	49.0	6	12.8
Age				
20–30	20	20.4	6	12.8
31–40	22	22.4	14	29.8
41–50	23	23.5	24	51.1
51–60	19	19.4	3	6.4
61 and above	14	14.3	0	0.0
average age	44		40	
Residency				
Less than 5 years	0	0.0	2	4.3
5–10 years	11	11.2	3	6.4
11–15 years	8	8.2	4	8.5
16–20 years	5	5.1	11	23.4
21 years or more	74	75.5	27	57.4
Education				
No formal education	11	11.2	0	0.0
Primary school	50	51.0	11	23.4
Secondary school	30	30.6	31	66.0
Certificate/Diploma	2	2.0	0	0.0
Degree holder	5	5.1	5	10.6
Occupation				
Fisherman	23	23.5	10	21.3
Farmer	5	5.1	0	0.0
Salaried individual	40	40.8	16	34.0
Unemployed	30	30.6	3	6.4
Did not answer			18	38.3

Meanwhile, in Karimunjawa, trends on the awareness level of mangroves and seagrasses' ES are the same (Figure 2). About 37.0% to 45.7% (mangroves) and 30.4% to 35.6% (seagrasses) of the respondents are "extremely aware" of supporting (serves a nursery, feeding, and breeding area), regulating (coastal protection and natural buffer), and cultural (recreational and educational) services. Also, 32.6% to 37.0% of the respondents are "very aware" that BCEs serve as habitats for many organisms and have water filtration functions. Interestingly, almost half (43.5% to 45.7%) of the respondents are "not aware" that these ecosystems are a great source of food. Moreover, similar patterns with Busuanga were observed in Karimunjawa for blue carbon functions, where 50.0% to 52.2% of the respondents are "moderate" to "extremely aware" while 47.8% to 50.0% are "slightly" to "not aware".

The Mann–Whitney U test was used to compare the awareness level between the two study sites (Table 2). Based on the analysis, the respondents in Karimunjawa have higher recognitions (M = 3.7 to 4.0, "very aware") of supporting (nursery, feeding, breeding area), regulating (coastal protection and natural buffer), and cultural (recreational and educational) services of mangroves than Busuanga's awareness level, which is "moderately aware" (M = 2.9 to 3.3). Conversely, provisioning services of mangroves (source of food) are perceived higher (M = 3.4, "moderately aware") in Busuanga compared to Karimunjawa (M = 2.6, "slightly aware"). For seagrass awareness, relatively similar trends with mangrove awareness were observed. Respondents in Karimunjawa have higher perceptions, "moderate" to "very aware" that seagrasses can serve as a nursery, feeding, and breeding ground (M = 3.7), habitat for many organisms (M = 3.7), protect coastal areas (M = 3.4), and act as a natural buffer (M = 3.5) compared to Busuanga's "slight" to "moderate" awareness (M = 2.4 to 3.0).

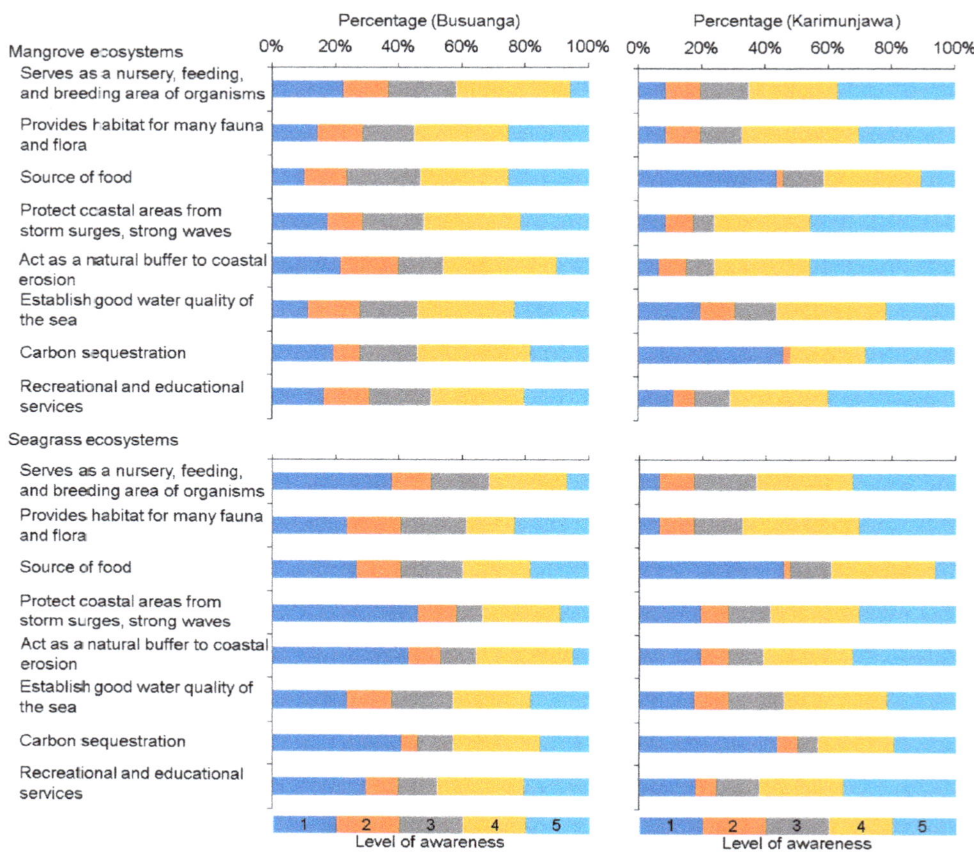

Figure 2. Awareness level of Busuanga (n = 98) and Karimunjawa (n = 47) respondents of BCE services. The awareness level is based on a five-point Likert scale (1 = not aware, 2 = slightly aware, 3 = moderately aware, 4 = very aware, 5 = extremely aware).

3.3. Respondents' Utilization of Blue Carbon Ecosystem Services

Generally, the provisioning and cultural services of BCEs are poorly utilized by the residents in both study sites (Figure 3). In Busuanga, 27.6% of the respondents have collected fishes and other seafood in mangrove areas at least once a week and 23.5% have done it once a month while the majority (33.7%) have never done it. About 25.5% of the respondents have collected seafood as an income source once a week, although a bigger portion (48.0%) have never utilized mangroves as an income source. Moreover, 75.5% of the sample size has never harvested mangroves as firewood materials. In terms of cultural services, only 22.4% of them have visited this habitat once a month for bird or bat watching while 75.5% have never utilized it for other recreational activities like paddling. Roughly 90.8% of the locals have never accessed the mangrove areas for research or educational purposes. Accessing seagrass meadows for its provisioning and cultural services has been observed to be very low, 53.1% to 92.9% of the respondents in Busuanga.

Table 2. Comparison of respondents' awareness level between Busuanga and Karimunjawa.

	Ecosystem Services	Busuanga (n = 98)		Karimunjawa (n = 47)	
		Weighted Mean [a]	Description	Weighted Mean [a]	Description
Mangrove ecosystems	Serves as a nursery, feeding, and breeding area of organisms	2.9 *	moderately aware	3.7 *	very aware
	Provides habitat for many fauna and flora	3.4	moderately aware	3.7	very aware
	Source of food	3.4 *	moderately aware	2.6 *	slightly aware
	Protect coastal areas from storm surges, strong waves	3.3 *	moderately aware	4.0 *	very aware
	Act as a natural buffer to coastal erosion	2.9 *	moderately aware	4.0 *	very aware
	Establish good water quality of the sea	3.4	moderately aware	3.3	moderately aware
	Carbon sequestration	3.3	moderately aware	2.9	moderately aware
	Recreational and educational services	3.2 *	moderately aware	3.8 *	very aware
Seagrass ecosystems	Serves as a nursery, feeding, and breeding area of organisms	2.5 *	slightly aware	3.7 *	very aware
	Provides habitat for many fauna and flora	3.0 *	moderately aware	3.7 *	very aware
	Source of food	2.9 **	moderately aware	2.5 **	slightly aware
	Protect coastal areas from storm surges, strong waves	2.4 *	slightly aware	3.4 *	moderately aware
	Act as a natural buffer to coastal erosion	2.4 *	slightly aware	3.5 *	very aware
	Establish good water quality of the sea	3.0	moderately aware	3.3	moderately aware
	Carbon sequestration	2.7	moderately aware	2.7	moderately aware
	Recreational and educational services	3.0	moderately aware	3.6	very aware

Notes: [a] Measured on a five-point Likert scale ranging from "not aware" (1) to "extremely aware" (5). Values with * ($p < 0.05$), ** ($p < 0.10$) are statistically different based on the Mann–Whitney U test.

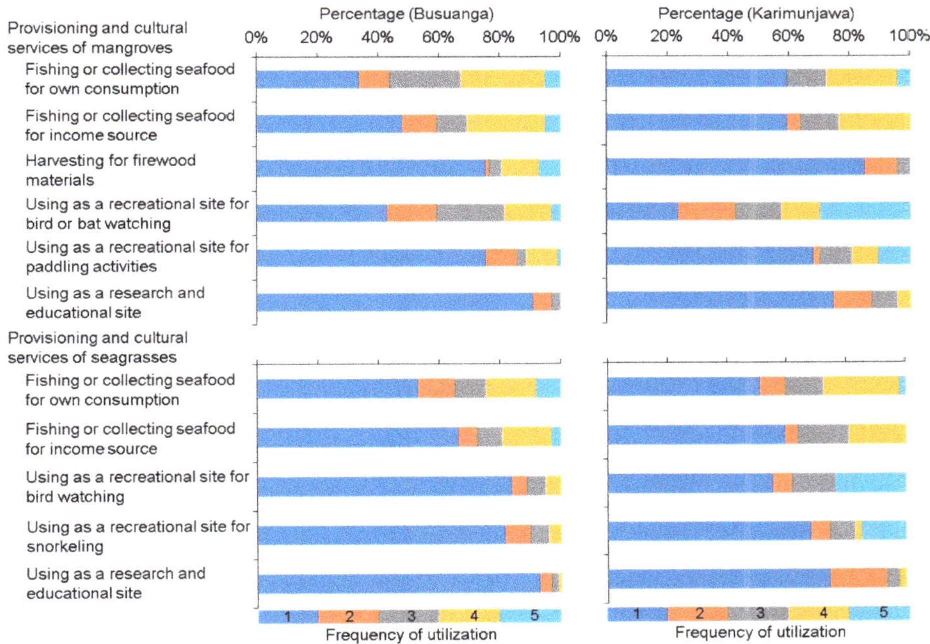

Figure 3. Utilization frequency of Busuanga (n = 98) and Karimunjawa (n = 47) respondents of provisioning and cultural services of BCEs. The frequency is based on a five-point Likert scale (1 = never use, 2 = once a year, 3 = once a month, 4 = once a week, 5 = every day).

Similar trends were observed in Karimunjawa, 23.4% of the respondents have collected fish and other seafood in mangroves once a week for their own consumption and income source while more than half (59.6%) have never utilized it (Figure 3). Harvesting mangroves for firewood materials was never done by 85.1% of the respondents while a few of them (10.6%) have harvested them at least once a year. For cultural services, bird or bat watching in mangrove areas is done every day by 29.8% of the sample size, 12.8% for once a week, 14.9% for once a month, and 19.1% for once a year while 23.4% have never done it. Meanwhile, 68.1% and 74.5% of the respondents have never accessed this habitat for paddling and educational purposes, respectively. Moreover, seagrass utilization is also low, more than half (51.1% to 74.5%) of the sample size has never accessed this ecosystem to collect seafood, as livelihood options, and as recreational sites. However, it is noteworthy that at least 25.5% of the respondents have gleaned in seagrass beds once a week and 23.4% of the respondents have observed birds in this habitat every day.

Although both utilization patterns in the study sites are relatively low, the comparison of weighted means through the Mann–Whitney U test showed significant differences in the results (Table 3). Higher values were obtained in Busuanga for using mangrove to collect food (M = 2.6) and firewood materials (M = 1.7) while Karimunjawa has a weighted mean of 2.1 and 1.2 for fishing (own consumption) and harvesting for firewood materials, respectively. Conversely, accessing seagrass beds for their cultural benefits is higher in Karimunjawa than in Busuanga. For instance, using this habitat as a site for bird watching and snorkeling activities are done at least once a year (M = 2.3 and M = 1.9, correspondingly) in the former while the respondents in the latter never used it for bird watching (M = 1.3) and snorkeling (M = 1.3).

Table 3. Comparison of respondents' utilization frequency of provisioning and cultural services between Busuanga and Karimunjawa.

	Provisioning and Cultural Services	Busuanga (n = 98)		Karimunjawa (n = 47)	
		Weighted Mean [a]	Description	Weighted Mean [a]	Description
Mangrove ecosystems	Fishing for own consumption	2.6 *	once a year	2.1 *	once a year
	Fishing for income source	2.3	once a year	2.0	once a year
	Harvesting for firewood materials	1.7 **	never use	1.2 **	never use
	Using as a recreational site (bird or bat watching)	2.2	once a year	3.1	once a month
	Using as a recreational site (paddling)	1.5	never use	1.9	once a year
	Used as a research or educational site	1.1	never use	1.4	never use
Seagrass ecosystems	Fishing for own consumption	2.1	once a year	2.2	once a year
	Fishing for income source	1.8	once a year	2.0	once a year
	Using as a recreational site (bird watching)	1.3 *	never use	2.3 *	once a year
	Using as a recreational site (snorkeling)	1.3 *	never use	1.9 *	once a year
	Used as a research or educational site	1.1 *	never use	1.3 *	never use

Notes: [a] Measured on a five-point Likert scale (1 = never use, 2 = once a year, 3 = once a month, 4 = once a week, 5 = everyday). Values with * ($p < 0.05$), ** ($p < 0.10$) are statistically different based on the Mann–Whitney U test.

3.4. Perceived Threats to Blue Carbon Ecosystems

The residents were asked to rank the threats based on their perceptions from most (1) to least (10) damaging threats (Table 4). Almost half (48.5%) of the respondents of Busuanga have identified natural disasters (i.e., typhoons, storm surges) to be the most (1st) concerning threat to mangroves whereas conversion to fishponds is perceived by 64.9% to be the least (10th) threat. Illegal cutting of mangroves (18.6%) came second in the rank along with pollution from domestic wastes (34.7%). Other perceived anthropogenic threats to mangroves like charcoal making, increasing population, informal settlers, building coastal infrastructures, conversion to nipa and coconut plantation, and conversion to residential areas ranked fourth (34.7%), fifth (30.6%), sixth (21.4%), seventh (26.8%), eighth (35.7%), and ninth (50.5%) places, respectively. Conversely, perceived threats in Karimunjawa are ranked differently (Table 4). For instance, 28.3% of the respondents of Karimunjawa have recognized natural disasters to be the least (10th) damaging threat to mangroves opposite to Busuanga's results. The difference is statistically significant ($p < 0.05$) based on the Mann–Whitney U test. Other significant differences on perceived threats observed in Karimunjawa include charcoal making (3rd, 21.7%), building infrastructures (third, 17.4%), conversion to fishponds (fifth, 19.6%), conversion to residential areas (sixth, 23.9%), conversion to palm tree plantation (eighth, 21.7%), and increasing population (ninth, 28.3%).

Table 4. Perceived threats to blue carbon ecosystems.

Perceived Threats	Busuanga			Karimunjawa		
Mangrove Ecosystems	**Mode [ab]**	**Percentage**	**Weighted Mean [b]**	**Mode [ab]**	**Percentage**	**Weighted Mean [b]**
Natural disasters	1	48.5	3.2 *	10	28.3	6.6 *
Pollution (domestic wastes)	2	34.7	2.8	2	23.9	3.7
Informal settlers	6	21.4	5.1	6	20.0	4.9
Increasing population	5	30.6	5.7 *	9	28.3	6.9 *
Charcoal making	4	34.7	3.6 *	3	21.7	5.5 *
Conversion to nipa and coconut/palm tree plantation	8	35.7	7.2 *	8	21.7	5.4 *
Mangrove cutting	2	18.6	3.9	1	26.1	4.7
Building infrastructures in coastal areas	7	26.8	6.5 *	3	17.4	5.5 *
Conversion to residential areas	9	50.5	7.7 *	6	23.9	5.6 *
Conversion to fishponds	10	64.9	8.9 *	5	19.6	5.1 *
Seagrass ecosystems						
Natural disasters	1	39.8	3.3 *	10	28.3	5.9 *
Pollution (domestic wastes)	1	43.9	2.1 *	1	28.3	4.1 *
Increasing population	3	22.4	5.5 *	6	21.7	6.7 *
Building infrastructures in coastal areas	4	17.3	5.7	6	21.7	5.0
Mangrove planting on seagrass beds	4	18.4	5.4	8	17.4	5.8
Sand mining	6	16.3	5.4	5	23.9	5.2
Unregulated gleaning	6	21.4	6.6 *	2	19.6	5.3 *
Siltation	6	17.3	5.5	2	17.4	5.2
Beach reclamation	9	33.7	7.4 *	4	19.6	4.6 *
Increasing sea surface temperature	10	52.0	8.4 *	10	23.9	6.0 *

Notes: [a] Most frequently occurring response, [b] measured from most (1) to least (10) damaging threats. * difference is statistically significant at $p < 0.05$ based on the Mann–Whitney U test.

Seagrass ecosystems are also highly vulnerable to natural and anthropogenic threats. Pollution from domestic wastes ranked first in the list of damaging threats in both study sites (Table 4). Similar to mangroves' threats, natural disasters occurring in Busuanga are also perceived by 39.8% of the sample size to be at the top list of threats whereas 28.3% of the respondents of Karimunjawa ranked it at the bottom of the list along with increasing sea surface temperature (23.9%). Moreover, mangrove planting in seagrass beds (18.4%) and building infrastructures in coastal areas (17.3%) in Busuanga tied in fourth place followed by sand mining (16.3%), unregulated gleaning (21.4%), and siltation (17.3%) all in sixth place. Beach reclamation and increasing sea surface temperature are perceived to be the least threats, occupying the 9th (33.7%) and 10th (52.0%) places, respectively. Meanwhile, in Karimunjawa, unregulated gleaning and beach reclamation are ranked significantly different ($p < 0.05$) taking the second (19.6%) and fourth (19.6) places, correspondingly.

3.5. Perceived Management Strategies to Blue Carbon Ecosystems

The respondents rated the strategies for BCEs management from the top (1) to least (7) priority measures. As shown in Table 5, 30.9% and 23.4% of the respondents in Busuanga have recognized "Organization strengthening and capacity development" and "Coastal and fisheries law enforcement" as the top strategies that need to be prioritized whereas

23.4% of the sample size in Karimunjawa acknowledged "Habitat management and marine sanctuaries" to be highly prioritized. Meanwhile, the least important strategies as perceived by the residents are "Information and educational campaigns" (29.0%) and "Coastal zoning" (27.7%) in Busuanga and "Coastal Zoning" (21.3%) in Karimunjawa.

Table 5. Perceived blue carbon ecosystem management strategies.

Perceived Management Strategies	Busuanga			Karimunjawa		
	Mode [ab]	Percentage	Weighted Mean [b]	Mode [ab]	Percentage	Weighted Mean [b]
Organization strengthening and capacity development	1	30.9	2.8 *	5	23.4	4.5 *
Coastal and Fisheries Law Enforcement	1	23.4	3.3	4	25.5	3.5
Fisheries Management	4	19.1	3.5	2	19.1	3.8
Habitat management and marine sanctuaries	3	22.3	4.1 *	1	23.4	3.5 *
Enterprise, livelihood, and tourism development	5	23.2	4.3	5	23.4	4.1
Information and educational campaigns	6	29.0	4.8 *	2	21.3	4.1 *
Coastal zoning	6	27.7	4.1	7	21.3	4.6

Notes: [a] Most frequently occurring response, [b] measured from the top (1) to least (7) priority management strategies. * difference is statistically significant at $p < 0.05$ based on the Mann–Whitney U test.

Moreover, residents were asked for their perceptions of who should be in charge of managing their BCEs (Figure 4). More than half (51.6% to 52.6%) of the respondents in Busuanga perceived that local residents should take the lead in the management of their mangroves and seagrasses while 21.7% to 23.9% of the sample size in Karimunjawa recognized that management should be a collective effort among local residents, local government, central government, NGOs and private sectors. Management by local government units is also acknowledged by 20.0% to 21.1% of the residents in Busuanga and 10.9% to 13.0% in Karimunjawa. A combined effort between local residents and local government units received fair recognition, 13.7% to 15.8%, in Busuanga and 15.2% in Karimunjawa. NGO-led managements are perceived by a small fraction, about 2.2% of the sample size in Karimunjawa whereas no recognition was recorded in Busuanga.

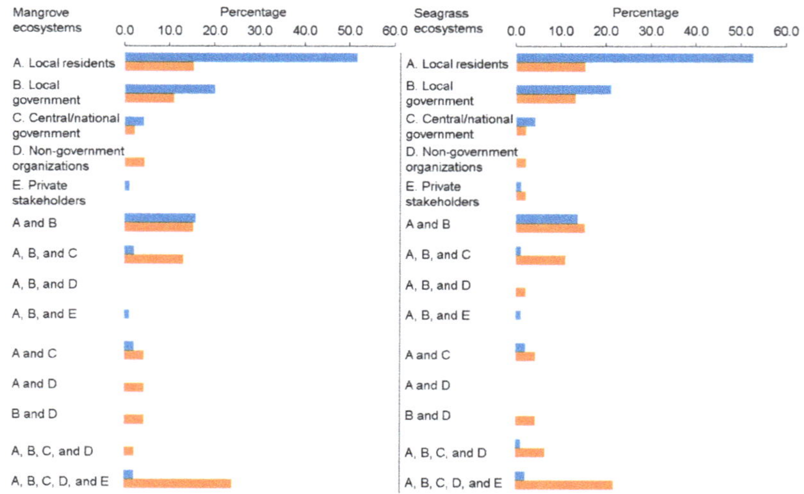

Figure 4. Perceived stakeholders to manage the blue carbon ecosystems. Blue bars represent Busuanga (n = 98) while orange bars reflect Karimunjawa (n = 47) respondents.

4. Discussions

4.1. Awareness and Utilization of Blue Carbon Ecosystems

The residents of Busuanga have a high recognition of mangroves because of their tangible benefits. Field observations and oral accounts documented that locals collect fish and shells in mangrove areas when their financial capacity to buy food is limited. This scenario is very common among coastal communities in the Philippines. Some locals are getting paid for assisting recreational activities such as firefly watching and paddling in mangrove areas. Moreover, coastal residents have first-hand experience with the protection services of mangroves when the super typhoon Haiyan hit the country in 2013. Ref. [33] have documented that mangrove services are highly recognized by the people when they are directly benefited by them; the more services they can get or observe, the higher they value the ecosystem. As for seagrass awareness, provisioning services are well acknowledged by the locals compared to other services (i.e., regulating, cultural). Unlike mangroves, seagrass ecosystem services are not well streamlined in coastal programs or often grouped with other ecosystems because the priorities for research and development activities are usually directed towards coastal resources with immediate economic impacts [34,55]. However, C3 Philippines, an NGO, is changing this trend. This group has already conducted several seagrass awareness campaigns (including blue carbon functions) on the island as part of their thrusts on the Dugong conservation program (program coordinator of C3 Philippines, personal communication, 19 July 2019).

The utilization frequency of BCE services in Busuanga is generally low despite residents' proximity to these resources (see Figure 3). Using correlation and multiple regressions, awareness level and sociodemographic characteristics of the locals were explored to determine whether these have an influence or none on their utilization behaviors (see Supplementary Materials). This study shows that the effect of social profile on the local's utilization behavior is inconclusive. As pointed out by [56], socio-demographics as predictor variables do not always influence their behavior. However, to some degree, the occupation of the residents influences their utilization behavior (as hypothesized); fishers are more active in using the BCEs services meanwhile others (i.e., farmers, employed individuals) are less active. Fishing in BCEs, though, is low in Busuanga since they utilize coral reefs more [57]. Harvesting mangroves for charcoal-making has reduced due to strict implementation of local ordinances and the establishment of marine protected areas on the island [57]. Moreover, cultural services offered by the BCEs are not used since most of the tourism-related activities on the island feature other ecosystems like coral reefs, beaches, and small islands.

The awareness level of BCE services in Karimunjawa reflects an overall low trend that agrees with previous perception studies on the island (i.e., [58,59]). Provisioning benefits are poorly received while cultural functions of BCEs are well perceived (see Figure 2). Regulating services like the blue carbon functions are generally unrecognized, while coastal protection and natural buffer services are acknowledged by the locals. To date, there are now different blue carbon opportunities in Indonesia [29], however, blue carbon related works are mainly focused on the carbon potential of BCEs (i.e., [60]) rather than on BCE awareness and capacity building among coastal communities.

Looking at the possible factors affecting their perceptions, the occupation of the locals is negatively associated with their awareness and utilization behaviors. Based on the 2017 population of Karimunjawa, fishers comprise 47.0% of the residents [45] thus, it is expected to document high utilization frequency (positive correlation). However, this study shows low utilization frequency (negative correlation); 59.6% to 68.1% of the locals do not perform fishing/gleaning activities. This may be due to the smaller number of fishers (21.3%) in the respondents and more (34.0%) salaried individuals. Accessibility to BCEs could potentially influence resident's usage rate [58]; however, despite the proximity of the respondents to BCEs, utilization yields are still low. This is because most of the fishers on the island are pelagic fishers [45]. Another factor that could influence their utilization is the level of awareness. Similar to Busuanga, the awareness and utilization in Karimunjawa correlate with each other. Residents who utilized the services (i.e., food source) are typically those who only recognized the benefits of BCEs. The work of [59] on the island also captured fewer fishing activities in mangroves because of their low awareness while gleaning activities in seagrass beds depend on the awareness of the abundance of associated organisms [58]. Interestingly, activities related to cultural activities are also low (see Figure 3) despite Karimunjawa becoming a popular tourism site [59].

4.2. Threats to Blue Carbon Ecosystems

Natural disturbances like typhoons resulting in strong waves and storm surges are perceived to be the most damaging threat to BCEs in Busuanga (see Table 4). The residents highly recognized natural calamities to destroy BCEs since they have personal experiences and observations. For example, when the super typhoon Haiyan devastated the Philippines in 2013, it caused significant damage to BCEs in the country [61,62]. During the conduct of household surveys, many residents recalled and shared their observations on how the super typhoon destroyed the mangroves and seagrasses. Similar findings were also documented from the residents in Eastern Samar where the super typhoon first hit the country [33].

Pollution from domestic wastes was also recognized to be one of the top concerning threats. Photo documents clearly showed solid wastes in coastal areas where communities live. The lack of discipline and effective solid waste management systems are common factors that propagate increasing pollution pressure to BCEs. Other concerning threats to mangroves include illegal harvesting for firewood and charcoal-making. This problem has been a consistent challenge to address since local communities have direct access to mangrove forests [1]; however, a recent survey in the locality has shown a decline in illegal activities due to the presence of local ordinances and national policies (i.e., [33]). Conversion to fishponds was identified as the least concerning threat since there are no converted aquaculture ponds in Busuanga. This is noteworthy since conversion to fishponds was one of the main causes of rapid mangrove degradation in the country in the early years [1]. Current programs and policies (i.e., Coastal Resource Management Programs, Revised Forestry Code of the Philippines (Presidential Decree No. 705)) at the local and national level have reduced these activities. Meanwhile, perceived threats specific to seagrasses include mangrove planting on seagrass beds, unregulated gleaning, and siltation. These pressures from human activities could result in a cascading effect in the whole coastal ecosystem. For instance, unregulated gleaning could result in biodiversity loss [63]. In the Philippines, cases of seagrass and associated organisms' decline were perceived and attributed to these human-induced stressors [55].

Conversely, the effect of natural disturbances on BCEs in Karimunjawa is perceived to be the least concerning threat as reflected in the results (see Table 4). Unlike the Philippines, the average number of tropical cyclones in Indonesia is seven (7) per year or about 9% of the average number of cyclones globally, because, in principle, tropical cyclones do not cross the country as they will always move away from the equator [64]. The lack of personal experiences and observations on the effect of natural disturbances on BCEs likely influenced residents' perceptions. In the perception study conducted by [59] in Karimunjawa, there are no questions about the effects of natural threats, as they are mostly human-derived. Their study showed that mangrove degradation in the island is correlated with human-derived disturbances; mangrove logging has the strongest correlation ($p < 0.01$) while the development of offshore inns (guesthouses) has the weakest correlation ($p < 0.05$). Conformingly, the results of this study also identified anthropogenic threats (i.e., mangrove cutting, pollution) as the top perceived causes of mangrove degradation. The recognition of human-induced threats to mangroves has been long supported by quantitative researches (i.e., [65,66]) and has been observed not just in Karimunjawa but also in other areas in Indonesia. For instance, mangrove losses due to conversion into oil palm plantation and other land uses were observed in the islands of Mentawau (West Sumatra), Lankat (North Sumatra), Bawal (West Kalimantan), Seram (Maluku), Bangka Belitung, and Enggano

Bengkulu [67]. The results of this study and previous researches (i.e., [59,68]) indicate that mangroves in Indonesia are still threatened by manmade activities despite an increase in management strategies (i.e., [66,69,70]).

Seagrasses in Karimunjawa are also perceived to be most threatened by anthropogenic disturbances (i.e., pollution, unregulated gleaning). There are not many published studies on seagrass ecosystems in Karimunjawa, however, seagrasses in small islands in Indonesia are generally highly vulnerable to human-induced activities [58]. Destructive fishing methods (push nets and trawls), coastal constructions, and sedimentation from coastal development are also among the top activities in Indonesia that damage the seagrasses [16]. Household wastes also affect the seagrasses; for example, in Spermonde Archipelago in South Sulawesi, seagrasses' health is influenced by the nutrient loading, turbidity, and total suspended solids coming from domestic solid and liquid wastes [71]. The results of this study conform with the survey conducted by [72] with seagrass experts in Indonesia such that current and future threats are mostly human-derived activities such as coastal development, sedimentation, poor water quality, seaweed farming, overexploitation of herbivores, and coastal erosion.

4.3. Management Strategies of Blue Carbon Ecosystems

In recent years, marine ecosystems in Busuanga have improved due to the presence of regulatory boards (i.e., PCSD), local authorities (i.e., MAO), and NGOs (i.e., C3 Philippines). There is also the creation of Fisheries and Aquatic Resources Management Councils (FARMCs) and different POs (i.e., fishers' association) which empowers communities to join management-related activities. However, despite the presence of different organizations, the residents prefer that organizational strengthening and capacity development and law enforcement should still be prioritized first in the list of management strategies (see Table 5). These perceptions agree with findings of [73] that community-based management has not been successful because communities lack self-sufficiency and their participation is merely rhetorical. There is a need to strengthen and capacitate communities in coastal management. Previous studies have documented that the ecosystems' (i.e., mangroves) conditions in Busuanga Island have improved through collaborative protective management with the POs [74] and a stronger presence of NGOs in the communities [73]. To further strengthen management strategies, FARMCs and POs have encouraged local constituents and members to actively participate in management-related activities such as coastal clean-up and mangrove planting. The positive reception of the locals in these initiatives could explain why around 50% of the respondents said locals should manage their BCEs (see Figure 4). Other areas in the Philippines have had practices where locals are active stakeholders in the management of BCEs [26].

The continued degradation of coastal resources on the island has also been linked to the weak presence and enforcement of habitat protection and management interventions [49]. Field observations and stories from the locals revealed that some illegal activities like mangrove cutting are still ongoing partly because of weak law enforcement. Ref. [44] documented the lack of strict implementation and law enforcement in their coastal governance and management system. There is also a concern for the ambiguity and overlapping roles of government organizations and NGOs, which can cause frustration and even conflict in the community [73]. Thus, it is important to establish different policies and plans for organizations in the community. For the lowest priority strategies, residents ranked information and educational campaigns and coastal zoning at the bottom (see Table 5) since these programs are already implemented on the island. Awareness campaigns are done by C3 Philippines, an NGO, for coastal communities around the island (C3 program coordinator, personal communications, 19 July 2020) while PCSD regulates the use of coastal zone [44].

The governance system in Karimunjawa has an overall weak performance in terms of addressing conflicts and meeting objectives to protect fishery resources from unsustainable and destructive practices [48]. The lack of stronger resource management on the island is reflected in the perceptions where locals recognized the need to prioritize the marine sanctuaries, habitat, and fisheries management (see Table 5). Almost half (47%) of the population on the island are fishers [45]; therefore, it is expected that residents prefer to prioritize management strategies affecting their livelihoods. Communities that have a high dependency on marine resources are generally more supportive of strategies related to fisheries management. There is also a need to carry out more information and educational campaigns to increase their awareness of BCEs. Although the results of this study show relatively high awareness of the benefits, the overall assessment reflects a lack of knowledge, which could result in weak participation in conservation activities (see Figure 4). For instance, the communities have relatively low motivation for participating in management-related activities because of poor awareness about the advantages of mangroves and the impacts of their degradation [59]. In addition, communities are not interested in joining the conservation activities due to a lack of direct economic incentives [75]. Moreover, the communities prefer that all stakeholders (i.e., residents, government, private sectors) should participate in managing their coastal resources (see Figure 4). This collaborative management approach on the island has long been implemented since 2007 [48]. In Indonesia, there are only a few cases of marine resource management that are co-managed by communities and the government [76]. The governance system should respect the customary knowledge, rules, decision-making process of the local communities to get the support of the communities [48]. Meanwhile, the lowest priority is coastal zoning. There is already a strict zonation policy that is being implemented and followed on the island [66] thus, the residents placed this strategy at the bottom of the list (see Table 5). The zonation of the island allows regulatory controls on different uses outlined in the management plan, like issuing permits to harvest some natural resources sustainably and conduct activities related to education and research purposes.

5. Implications to Management of Blue Carbon Ecosystems in the Coral Triangle Region

Despite the gaining momentum of the "blue carbon" discourse and collaborative action in the international and national arenas, there is still a lot of work to be done at the local level, particularly where local governments and implementers are the ones interacting with these ecosystems. This study presents an opportunity to level off BCEs perceptions at the local level in the CTR, particularly in the countries of Indonesia and the Philippines. People's awareness level, utilization rates, and perceived threats and management strategies are used as proxies to determine the familiarity of coastal communities with BCEs. The results of this work show that local communities are aware of the services they can get from BCEs. However, their awareness depends entirely on the type of benefit they directly receive. For instance, provisioning services (i.e., food source) is fairly acknowledged in Busuanga while poorly recognized in Karimunjawa. Personal experiences of the residents also greatly influence their perception of the BCEs.

Factors affecting their perceptions were explored in this study as well. Although the effect of socio-demographics on awareness and utilization in this work is inconclusive, it is important to consider these factors when looking at the role of communities in coastal management. Not only the quantitative aspects of the effect but qualitative aspects can be analyzed applying text mining methods to the interview results of locals [77,78]. The willingness of the locals to actively participate in management-related activities depends on their awareness level. In Busuanga, where locals have high regard for BCEs, they are willing to manage them. In contrast, residents in Karimunjawa prefer their local government or a multisectoral management scheme since they have low comprehension of the BCE services. This observation is a useful indicator in strategizing advocacy campaigns and the levels of engagement of local stakeholders in the CTR. It is noteworthy that engaging local communities in ecosystem service assessments helps define their role in multi-governance of the environments as well as the importance of ES and the factors that influence social preferences and trade-offs related to land-use change and decision-making [38].

Another implication of this study on BCEs in the CTR is the perceived threats. To enable the sustainable management of these resources in the region, threats that destroy them should be identified first. Damages caused by natural disturbances are hard to address but restoration and rehabilitation of BCEs after a catastrophic event can be done [1,55] while anthropogenic threats are the ones that can be prevented and addressed directly. This study documented that human activities (i.e., cutting, coastal development) are concerning threats to BCEs in Busuanga and Karimunjawa. Through the locals' responses, different stakeholders can have an opportunity to address present and future threats at local scales where management strategies are often weak compared to the national level. Reducing or prohibiting these activities can ensure the proliferation of BCEs in the CTR.

This perception study, particularly on the section where locals were asked to prioritize management efforts, to some extent, served as (a) a feedback mechanism on the impact of prior and/or existing BCE management activities; and (b) an assessment tool that helps identify the gaps of the management plans and programs for the two countries. First, it was made evident that in both sites, the least prioritized management actions are the ones that are already being strongly and widely, if not effectively, implemented such as information and educational campaigns in Busuanga and coastal zoning in Karimunjawa. Second, the locals' responses collectively revealed that certain management activities, like the ones perceived to be prioritized—organization strengthening and capacity building in Busuanga and habitat management and fish sanctuaries in Karimunjawa—are the activities or programs that are less felt and experienced. This study is a crucial learning in how collective perceptions can be used moving forward in BCE management strategies in each country as well as a tool for identifying common grounds that can be shared and are transferrable across the CTR.

Supplementary Materials: The following are available online at https://www.mdpi.com/2071-1050/13/1/127/s1, Texts S1: Factors influencing the awareness and utilization of blue carbon ecosystem services, Table S1: Multiple linear regressions of socio-demographic profile and awareness level of (S1a) mangroves and (S1b) seagrasses' ecosystem services, Table S2: Multiple linear regressions of socio-demographic profile and utilization frequency of (S2a) mangroves and (S2b) seagrasses' provisioning and cultural services, Table S3: Correlation analysis of locals' awareness and utilization of (S3a) mangroves and (S3b) seagrasses' ecosystem services.

Author Contributions: Conceptualization, J.M.D.Q., Y.U., and R.K.; methodology, J.M.D.Q., K.M.L., and Y.U.; validation, J.M.D.Q., K.M.L., and Y.U.; formal analysis, J.M.D.Q. and Y.U., investigation, J.M.D.Q., Y.U., and K.M.L.; resources, J.M.D.Q., K.M.L., and Y.U.; writing—original draft preparation, J.M.D.Q.; writing—review and editing, J.M.D.Q., Y.U., K.M.L., and R.K.; supervision, Y.U. and R.K.; project administration. Y.U. and R.K.; supplementary materials, J.M.D.Q.; funding acquisition, R.K. All authors have read and agreed to the published version of the manuscript.

Funding: This research was funded by JSPS KAKENHI, grant numbers JP20K12398, JP16KK0053, JP17K02105, JP17H01682, Japan Science and Technology Agency (JST) and Japan International Cooperation Agency (JICA) through the Science and Technology Research Partnership for Sustainable Development Program (SATREPS)—Comprehensive Assessment and Conservation of Blue Carbon Ecosystems and Their Services in the Coral Triangle (Blue CARES) project; Toyota Foundation (D17-N-0107); Foundation for Environmental Conservation Measures, Keidanren (2020); Kurita Water and Environment Foundation (20C002).

Informed Consent Statement: Informed consent was obtained from all subjects involved in the study.

Data Availability Statement: The data presented in this study are available on request from the corresponding author. The data are not publicly available since this study is part of an on-going research project.

Acknowledgments: The authors would like to thank the key informants from different government units of Busuanga, Philippines and Karimunjawa National Park, Indonesia for their support in the conduct of the interviews.

Conflicts of Interest: The authors declare that they have no known competing financial interests or personal relationships that could have appeared to influence the work reported in this paper.

References

1. Primavera, J.H. Development and Conservation of Philippine Mangroves: Institutional Issues. *Ecol. Econ.* **2008**, *35*, 91–106. [CrossRef]
2. Alongi, D.M. Mangrove forests: Resilience, protection from tsunamis, and responses to global climate change. *Estuar. Coast. Shelf Sci.* **2008**, *76*, 1–13. [CrossRef]
3. Mukherjee, N.; Sutherland, W.J.; Dicks, L.; Hugé, J.; Koedam, N.; Dahdouh-Guebas, F. Ecosystem Service Valuations of Mangrove Ecosystems to Inform Decision Making and Future Valuation Exercises. *PLoS ONE* **2014**, *9*, e111386. [CrossRef] [PubMed]
4. Dasgupta, S.; Islam, S.; Huq, M.; Khan, Z.H.; Hasib, R. Quantifying the protective capacity of mangroves from storm surges in coastal Bangladesh. *PLoS ONE* **2019**, *14*, e0214079. [CrossRef] [PubMed]
5. Uddin, S.; de Ruyter van Steveninck, E.; Stuip, M.; Shah, M.A.R. Economic valuation of provisioning and cultural services of a protected mangrove ecosystem: A case study on Sundarbans Reserve Forest, Bangladesh. *Ecosyst. Serv.* **2013**, *5*, 88–93. [CrossRef]
6. Crooks, S.; von Unger, M.; Schile, L.; Allen, C.; Whisnant, R. *Understanding Strategic Blue Carbon Opportunities in the Seas of East Asia*; Report by Silvestrum Climate Associates for Partnerships in Environmental Management for the Seas of East Asia (PEMSEA), Conservation International and The Nature Conservancy, with support from the Global Environment Facility and United Nations Development Program; Partnerships in Environmental Management for the Seas of East Asia (PEMSEA): Quezon City, Philippines, 2017.
7. Howard, J.; Sutton-Grier, A.; Herr, D.; Kleypas, J.; Landis, E.; McLeod, E.; Pidgeon, E.; Simpson, S. Clarifying the role of coastal and marine systems in climate mitigation. *Front. Ecol. Environ.* **2017**, *15*, 42–50. [CrossRef]
8. Pidgeon, E. Carbon sequestration by coastal marine habitats: Important missing sinks. In *The Management of Natural Coastal Carbon Sinks*; Laffoley, D., Grimsditch, G., Eds.; IUCN: Gland, Switzerland, 2009; p. 53.
9. McLeod, E.; Chmura, G.L.; Bouillon, S.; Salm, R.; Björk, M.; Duarte, C.M.; Lovelock, C.E.; Schlesinger, W.H.; Silliman, B.R. A blueprint for blue carbon: Toward an improved understanding of the role of vegetated coastal habitats in sequestering CO_2. *Front. Ecol. Environ.* **2011**, *9*, 552–560. [CrossRef]
10. Nellemann, C.; Corcoran, E.; Duarte, C.M.; Valdrés, L.; Young, C.D.; Fonseca, L.; Grimsditch, G. *Blue Carbon—The Role of Healthy Oceans in Binding Carbon*; UN Environment, GRID-Arendal: Arendal, Norway, 2009.
11. Beaumont, N.; Jones, L.; Garbutt, A.; Hansom, J.; Toberman, M. The value of carbon sequestration and storage in coastal habitats. *Estuar. Coast. Shelf Sci.* **2014**, *137*, 32–40. [CrossRef]
12. Asaad, I.; Lundquist, C.; Erdmann, M.V.; Costello, M.J. Delineating priority areas for marine biodiversity conservation in the Coral Triangle. *Biol. Conserv.* **2018**, *222*, 198–211. [CrossRef]
13. Giri, C.; Ochieng, E.; Tieszen, L.L.; Zhu, Z.; Singh, A.K.; Loveland, T.R.; Masek, J.G.; Duke, N.C. Status and distribution of mangrove forests of the world using earth observation satellite data. *Glob. Ecol. Biogeogr.* **2010**, *20*, 154–159. [CrossRef]
14. Alongi, D.M.; Murdiyarso, D.; Fourqurean, J.W.; Kauffman, J.B.; Hutahaean, A.; Crooks, S.; Lovelock, C.E.; Howard, J.B.; Herr, D.; Fortes, M.D.; et al. Indonesia's blue carbon: A globally significant and vulnerable sink for seagrass and mangrove carbon. *Wetl. Ecol. Manag.* **2015**, *24*, 3–13. [CrossRef]
15. Long, J.B.; Giri, C. Mapping the Philippines' Mangrove Forests Using Landsat Imagery. *Sensors* **2011**, *11*, 2972–2981. [CrossRef] [PubMed]
16. UNEP. *National Reports on Seagrass in the South China Sea*; UNEP/GEF/SCS Technical Publication No. 12; United Nations Environment Programme: Bangkok, Thailand, 2008.
17. Greenberg, R.; Maldonado, J.E.; Droege, S.; McDonald, M.V. Tidal Marshes: A Global Perspective on the Evolution and Conservation of Their Terrestrial Vertebrates. *BioScience* **2006**, *56*, 675. [CrossRef]
18. Chmura, G.L.; Anisfeld, S.C.; Cahoon, D.R.; Lynch, J.C. Global carbon sequestration in tidal, saline wetland soils. *Glob. Biogeochem. Cycles* **2003**, *17*, 1111. [CrossRef]
19. Munang, R.; Thiaw, I.; Rivington, M. Ecosystem Management: Tomorrow's Approach to Enhancing Food Security under a Changing Climate. *Sustainability* **2011**, *3*, 937–954. [CrossRef]
20. Costanza, R.; de Groot, R.; Sutton, P.; van der Ploeg, S.; Anderson, S.J.; Kubiszewski, I.; Farber, S.; Turner, R.K. Changes in the global value of ecosystem services. *Glob. Environ. Chang.* **2014**, *26*, 152–158. [CrossRef]
21. Spalding, M.; Parrett, C.L. Global patterns in mangrove recreation and tourism. *Mar. Policy* **2019**, *110*, 103540. [CrossRef]
22. Crooks, S.; Herr, D.; Tamelander, J.; Laffoley, D.; Vandever, J. Mitigating climate change through restoration and management of coastal wetlands and near-shore marine ecosystems: Challenges and opportunities. In *Environment Department Papers, No. 121. Marine Ecosystem Series*; World Bank: Washington, DC, USA, 2011.
23. Fourqurean, J.W.; Duarte, C.M.; Kennedy, H.; Marbà, N.; Holmer, M.; Mateo, M.A.; Apostolaki, E.T.; Kendrick, G.A.; Krause-Jensen, D.; McGlathery, K.J.; et al. Seagrass ecosystems as a globally significant carbon stock. *Nat. Geosci.* **2012**, *5*, 505–509. [CrossRef]
24. Duarte, C.M.; Losada, I.J.; Hendriks, I.E.; Mazarrasa, I.; Marbà, N. The role of coastal plant communities for climate change mitigation and adaptation. *Nat. Clim. Chang.* **2013**, *3*, 961–968. [CrossRef]
25. Murdiyarso, D.; Purbopuspito, J.; Kauffman, J.B.; Warren, M.W.; Sasmito, S.D.; Donato, D.C.; Manuri, S.; Krisnawati, H.; Taberima, S.; Kurnianto, S. The potential of Indonesian mangrove forests for global climate change mitigation. *Nat. Clim. Chang.* **2015**, *5*, 1089–1092. [CrossRef]

26. Gevaña, D.T.; Pulhin, J.M.; Tapia, M.A. Chapter 13—Fostering climate change mitigation through a community-based approach: Carbon stock potential of community-managed mangroves in the Philippines. In *Coastal Management*; Krishnamurthy, R.R., Jonathan, M.P., Srinivasalu, S., Glaeser, B., Eds.; Academic Press: London, UK, 2019; pp. 271–282.
27. Wahyudi, A.J.; Rahmawati, S.; Irawan, A.; Hadiyanto, H.; Prayudha, B.; Hafizt, M.; Afdal, A.; Adi, N.S.; Rustam, A.; Hernawan, U.E.; et al. Assessing Carbon Stock and Sequestration of the Tropical Seagrass Meadows in Indonesia. *Ocean Sci. J.* **2020**, *55*, 85–97. [CrossRef]
28. Thompson, B.S.; Primavera, J.H.; Friess, D.A. Governance and implementation challengs for mangrove forest Payments for Ecosystem Services (PES): Empirical evidence from the Philippines. *Ecosyst. Serv.* **2017**, *23*, 146–155. [CrossRef]
29. Murdiyarso, D.; Sukara, E.; Suprianta, J.; Koropitan, A.; Juliandi, B.; Jompa, J. *Creating Blue Carbon Opportunities in the Maritime Archipelago Indonesia*; CIFOR Policy Brief No. 3; CIFOR: Bogor, Indonesia, 2018. [CrossRef]
30. Lukman, K.M.; Quevedo, J.M.D.; Kakinuma, K.; Uchiyama, Y.; Kohsaka, R. Indonesia Provincial Spatial Plans on mangroves in era of decentralization: Application of content analysis to 27 provinces and "blue carbon" as overlooked components. *J. For. Res.* **2019**, *24*, 341–348. [CrossRef]
31. Herr, D.; Blum, J.; Himes-Cornell, A.; Sutton-Grier, A.E. An analysis of the potential positive and negative livelihood impacts of coastal carbon offset projects. *J. Environ. Manag.* **2019**, *235*, 463–479. [CrossRef] [PubMed]
32. Lukman, K.M.; Uchiyama, Y.; Quevedo, J.M.D.; Kohsaka, R. Local awareness as an instrument for management and conservation of seagrass ecosystem: Case of Berau Regency, Indonesia. *Ocean Coast. Manag.* **2020**, 105451. [CrossRef]
33. Quevedo, J.M.D.; Uchiyama, Y.; Kohsaka, R. Perceptions of local communities on mangrove forests, their services and management: Implications for Eco-DRR and blue carbon management for Eastern Samar, Philippines. *J. For. Res.* **2019**, *25*, 1–11. [CrossRef]
34. Quevedo, J.M.D.; Uchiyama, Y.; Kohsaka, R. Perceptions of the seagrass ecosystems for the local communities of Eastern Samar, Philippines: Preliminary results and prospects of blue carbon services. *Ocean Coast. Manag.* **2020**, *191*, 105181. [CrossRef]
35. Coral Triangle Initiative Climate Change Adaptation Working Group. Regional Workshop on Blue Carbon. *Workshop Proceedings Report*. 2017. Available online: http://www.coraltriangleinitiative.org/sites/default/files/resources/Regional_CTI_BC_Workshop_Report.pdf (accessed on 19 July 2020).
36. International Partnership for Blue Carbon (IPBC). Draft Strategic Plan. 2016. Available online: https://bluecarbonpartnership.org/wp-content/uploads/2016/11/IPBC-Strategic-Plan-October-2016.pdf (accessed on 19 July 2020).
37. Quintas-Soriano, C.; Brandt, J.S.; Running, K.; Baxter, C.V.; Gibson, D.M.; Narducci, J.; Castro, A.J. Social-ecological systems influence ecosystem service perception: A Programme on Ecosystem Change and Society (PECS) analysis. *Ecol. Soc.* **2017**, *23*. [CrossRef]
38. Millennium Ecosystem Assessment. *Ecosystems and Human Well-Being (MEA)*; Island Press: Washington, DC, USA, 2005; pp. 25–36.
39. Díaz, S.; Demissew, S.; Carabias, J.; Joly, C.A.; Lonsdale, M.; Ash, N.; Larigauderie, A.; Adhikari, J.R.; Arico, S.; Báldi, A.; et al. The IPBES Conceptual Framework—Connecting nature and people. *Curr. Opin. Environ. Sustain.* **2015**, *14*, 1–16. [CrossRef]
40. Kovács, B.; Uchiyama, Y.; Miyake, Y.; Penker, M.; Kohsaka, R. An explorative analysis of landscape value perceptions of naturally dead and cut wood: A case study of visitors to Kaisho Forest, Aichi, Japan. *J. For. Res.* **2020**, *25*, 291–298. [CrossRef]
41. Uchiyama, Y.; Kohsaka, R. Cognitive value of tourism resources and their relationship with accessibility: A case of Noto region, Japan. *Tour. Manag. Perspect.* **2016**, *19*, 61–68. [CrossRef]
42. Kohsaka, R. Developing biodiversity indicators for cities: Applying the DPSIR model to Nagoya and integrating social and ecological aspects. *Ecol. Res.* **2010**, *25*, 925–936. [CrossRef]
43. Ehara, M.; Hyakumura, K.; Sato, R.; Kurosawa, K.; Araya, K.; Sokh, H.; Kohsaka, R. Addressing Maladaptive Coping Strategies of Local Communities to Changes in Ecosystem Service Provisions Using the DPSIR Framework. *Ecol. Econ.* **2018**, *149*, 226–238. [CrossRef]
44. Bautista, M.A.; Malolos, G.A.; Magyaya, A.; Palevino, M.L.; Suarez, M.A. Municipality of Busuanga ECAN Resource Management Plan 2017–2022. In Partnership with Municipal Government of Busuanga, Municipal ECAN Board, Palawan Council for Sustainable Development. Available online: https://pcsd.gov.ph/2020/12/11/07-busuanga-ecan-resource-management-plan/ (accessed on 24 December 2020).
45. Hafsaridewi, R.; Fahrudin, A.; Sutrisno, D.; Koeshendrajana, S. Resource management in the Karimunjawa Islands, Central Jawa of Indonesia, through DPSIR approach. *Adv. Environ. Sci.* **2018**, *10*, 7–22.
46. Prasetya, J.D.; Ambariyanto, A.; Supriharyono; Purwanti, F. Mangrove Health Index as Part of Sustainable Management in Mangrove Ecosystem at Karimunjawa National Marine Park Indonesia. *Adv. Sci. Lett.* **2017**, *23*, 3277–3282. [CrossRef]
47. Ambal, R.G.R.; Duya, M.V.; Cruz, M.A.; Coroza, O.G.; Vergara, S.G.; de Silva, N.; Molinyawe, N.; Tabaranza, B. Key biodiversity areas in the Philippines: Priorities for conservation. *J. Threat. Taxa* **2012**, *4*, 2788–2796. [CrossRef]
48. Campbell, S.J.; Kartawijaya, T.; Yulianto, I.; Prasetia, R.; Clifton, J. Co-management approaches and incentives improve management effectiveness in the Karimunjawa National Park, Indonesia. *Mar. Policy* **2013**, *41*, 72–79. [CrossRef]
49. Magbanua, F.; D'agnes, L.; Castro, J. *Integrated Coastal Management Makes a Difference to Human and Ecosystem Health: Evidence from Philippines*; IPOPCORM Monograph Series No. 4; PATH Foundation Philippines Inc: Makati City, Philippines, 2007.
50. Purwanti, F.; Alikodra, H.S.; Basuni, S.; Soedharma, D. Pengembangan Co-Management Taman Nasional Karimunjawa [Co-Management Improvement Karimunjawa National Park]. *Ilmu Kelautan* **2008**, *13*, 159–166.

51. Sarmah, H.K.; Hazarika, B.B. Importance of the size of sample and its determination in the context of data related to the schools of greater Guwahati. *Bull. Gauhati Univ. Math. Assoc.* **2012**, *12*, 55–76.
52. Sarmah, H.K.; Hazarika, B.B.; Choudhury, G. An investigation on effect of bias on determination of sample size on the basis of data related to the students of schools of Guwahati. *Int. J. Appl. Math. Stat. Sci.* **2013**, *2*, 33–48.
53. Bartlett, J.E., II.; Kotrlik, J.W.; Higgins, C.C. Organizational research: Determining appropriate sample size in survey research. *Inf. Technol. Learn. Perform. J.* **2001**, *19*, 43–50.
54. Philippine Statistics Authority. 2015 Census of Population, Report No. 2—Demographic and Socioeconomic Characteristics Palawan. 2017. Available online: https://psa.gov.ph/sites/default/files/17_Palawan.pdf (accessed on 13 December 2020).
55. Fortes, M.D. A review: Biodiversity, distribution and conservation of Philippine seagrasses. *Philipp. J. Sci.* **2013**, *142*, 95–111.
56. McGehee, N.; Andereck, K.L. Factors predicting rural residents' support of tourism. *J. Travel Res.* **2004**, *43*, 131–140. [CrossRef]
57. Garces, L.; Pido, M.D.; Tupper, M.H.; Silvestre, G. Evaluating the management effectiveness of three marine protected areas in the Calamianes Islands, Palawan Province, Philippines: Process, selected results and their implications for planning and management. *Ocean Coast. Manag.* **2013**, *81*, 49–57. [CrossRef]
58. Nurdin, N.; Riani, E.; Djuwita, I.; Budiharsono, S.; Purbayanto, A.; Asmus, H. Challenging for seagrass management in Indonesia. *J. Coast. Dev.* **2012**, *15*, 234–242.
59. Puryono, S.; Suryanti, S. Degradation of mangrove ecosystem in Karimunjawa Island based on public perception and management. In Proceedings of the IOP Conference Series: Earth and Environmental Science, Semarang, Indonesia, 30–31 October 2018; Volume 246, p. 012080. [CrossRef]
60. Nehren, U.; Wicaksono, P. Mapping soil carbon stocks in an oceanic mangrove ecosystem in Karimunjawa Islands, Indonesia. *Estuar. Coast. Shelf Sci.* **2018**, *214*, 185–193. [CrossRef]
61. Alura, D.P.; Alura, N.C.; Alura, R.P.C. Mangrove forest and seagrass bed of Eastern Samar, Philippines: Extent of damage by typhoon Yolanda. *IJNRLS* **2015**, *2*, 30–35.
62. Villamayor, B.M.R.; Rollon, R.N.; Samson, M.S.; Albano, G.M.G.; Primavera, J.H. Impact of Haiyan on Philippine mangroves: Implications to the fate of the widespread monospecific Rhizophora plantations against strong typhoons. *Ocean Coast. Manag.* **2016**, *132*, 1–14. [CrossRef]
63. Nordlund, L.M.; Gullström, M. Biodiversity loss in seagrass meadows due to local invertebrate fisheries and harbor activities. *Estuar. Coast. Shelf Sci.* **2013**, *135*, 231–240. [CrossRef]
64. Mulyana, E.; Prayoga, M.B.R.; Yananto, A.; Wirahma, S.; Aldrian, E.; Harsoyo, B.; Seto, T.H.; Sunarya, Y. Tropical cyclones characteristics in Southern Indonesia and the impact of extremes rainfall event. In Proceedings of the MATEC Web of Conferences ICDM, Andalas University, Padang, Indonesia, 2–4 May 2018; Volume 229, pp. 1–79. [CrossRef]
65. Kamal, M.; Harjo, H.; Wicaksono, P.; Adi, N.S.; Arjasakusuma, S. Assessment of mangrove forest degradation through canopy fractional cover in Karimunjawa Island, Central Java, Indonesa. *Geoplanning J. Geomat. Plan.* **2016**, *3*, 107–116. [CrossRef]
66. Ariyani, N.A.E. Implementation of conservation policy through the protection of life support system in the Karimunjawa National Park. In Proceedings of the E3S Web of Conferences, Semarang, Indonesia, 15–16 August 2018; Volume 31. [CrossRef]
67. Kotijah, S.; Ventyrina, I. Preventive regulations to remove environmental damage to mangrove ecosystem in East Kalimantan, Indonesia. *Int. J. Res. Law Econ. Soc. Sci.* **2019**, *1*, 9–19. [CrossRef]
68. Margono, B.A.; Potapov, P.V.; Turubanova, S.; Stolle, F.; Hansen, M.C. Primary forest cover loss in Indonesia over 2000–2012. *Nat. Clim. Chang.* **2014**, *4*, 730–735. [CrossRef]
69. Armitage, D. Socio-institutional dynamics and the political ecology of mangrove forest conservation in Central Sulawesi, Indonesia. *Glob. Environ. Chang.* **2002**, *12*, 203–217. [CrossRef]
70. UNEP. *Importance of Mangroves to People: A Call to Action*; van Bochove, J., Sullivan, E., Nakamura, T., Eds.; United Nations Environment Programme World Conservation Monitoring Centre: Cambridge, UK, 2014; 128p.
71. Amri, L.; Setiadi, D.; Qayim, I.; Djokosetiyanto, D. Nutrient content of seagrass Enhalus acoroides leaves in Barranglompo and Bonebatang Islands: Implication to increased anthropogenic pressure. *Indones. J. Mar. Sci.* **2011**, *16*, 181–186. [CrossRef]
72. Unsworth, R.K.F.; Ambo-Rappe, R.; Jones, B.L.; La Nafie, Y.A.; Irawan, A.; Hernawan, U.E.; Moore, A.M.; Cullen-Unsworth, L.C. Indonesia's globally significant seagrass meadows are under widespread threat. *Sci. Total. Environ.* **2018**, *634*, 279–286. [CrossRef]
73. Austin, R.L.; Eder, J.F. Environmentalism, Development, and Participation on Palawan Island, Philippines. *Soc. Nat. Resour.* **2007**, *20*, 363–371. [CrossRef]
74. D'Agnes, L.; D'Agnes, H.E.; Schwartz, J.B.; Amarillo, M.L.; Castro, J. Integrated management of coastal resources and human health yields added value: A comparative study in Palawan. *Environ. Conserv.* **2010**, *37*, 398–409. [CrossRef]
75. Abdullah, K.; Said, A.M.; Omar, D. Community-based Conservation in Managing Mangrove Rehabilitation in Perak and Selangor. *Procedia Soc. Behav. Sci.* **2014**, *153*, 121–131. [CrossRef]
76. Glaser, M.; Baitoningsih, W.; Ferse, S.C.; Neil, M.; Deswandi, R. Whose sustainability? Top–down participation and emergent rules in marine protected area management in Indonesia. *Mar. Policy* **2010**, *34*, 1215–1225. [CrossRef]
77. Kohsaka, R.; Matsuoka, H. Analysis of Japanese municipalities with Geopark, MAB, and GIAHS certification: Quantitative approach to official records with text-mining methods. *SAGE Open* **2015**, *5*, 1–10. [CrossRef]
78. Kohsaka, R.; Matsuoka, H.; Uchiyama, Y.; Rogel, M. Regional management and biodiversity conservation in GIAHS: Text analysis of municipal strategy and tourism management. *Ecosyst. Health Sustain.* **2019**, *5*, 124–132. [CrossRef]

Article

Importance of Urban Green at Reduction of Particulate Matters in Sihwa Industrial Complex, Korea

Sin-Yee Yoo, Taehee Kim, Suhan Ham, Sumin Choi and Chan-Ryul Park *

Urban Forests Research Center, National Institute of Forest Science, 57, Hoegiro, Dongdaemun-gu, Seoul 02455, Korea; dbtlsdl94@korea.kr (S.-Y.Y.); kth9064@korea.kr (T.K.); hamsu2498@si.re.kr (S.H.); ciromi@korea.kr (S.C.)

* Correspondence: maeulsoop@korea.kr; Tel.: +82-2-961-2612

Received: 5 July 2020; Accepted: 12 September 2020; Published: 16 September 2020

Abstract: The utilization of urban green areas has increased, but it is unclear whether urban green areas can decrease the concentration of particulate matter at an industrial complex city in Korea. We measured the extent of particulate matter (PM) reduction at a buffer green area in the Sihwa Industrial Complex. PM was measured at the industrial complex, the urban green area, and a nearby residential area from April to October 2019. PM reduction rates were highest at the urban green area in August and October, which is related to increased atmospheric mixing height and the active west wind blowing from the industrial complex to the residential area. Reduction rates of PM_{10} and $PM_{2.5}$ at the urban green area showed the lowest values, namely 14.4% and 25.3%, respectively. The air temperature, wind speed, and humidity could affect the PM reduction rate by influencing the movement and dispersion of PM at the micro-spatiotemporal scale. These results indicated that PM concentration could be reduced by the structural change of a forest layer at a micro scale in urban green areas.

Keywords: urban green area; Sihwa Industrial Complex; air pollutants; PM reduction; atmospheric mixing height; wind speed

1. Introduction

High domestic PM concentration is known to lead to increased human activity at urban green areas in industrial complex cities. The PM concentration in Korea is usually expressed as the same concentration according to the area where the national atmospheric monitoring station is installed. However, the actual PM values vary, and they change in accordance with the existence of the urban green areas [1,2]. Thus, citizens want to know information about the PM concentration under the forest canopy at hiking time. Urban green area is currently regarded as an important PM reduction measure that is highly accessible to citizens, and it also performs the role of atmospheric purification [3–5]. As a result, urban green area is widely used to manage the urban atmosphere [6,7].

Urban green area can reduce PM by absorption, blocking, and deposition. In the process of photosynthesis, the trees in the green area can absorb gaseous air pollutants such as carbon dioxide (CO_2), nitrogen dioxide (NO_2), and sulfur dioxide (SO_2) through the stomata of the leaves [8]. The tree canopy can also reduce the speed and movement of PM, resulting in PM being blocked and then falling to the surface of the ground [9,10]. Areas with large forest cover ratios were found to have lower PM concentrations than highly urbanized areas [2,7]. However, some studies have reported that the effects of urban green area on the urban atmosphere appear to be limited. The presence of roadside green area interferes with the airflow, causing roadside PM to stagnate and resulting in a higher

PM concentration at the roadside green area than at other areas on the roadside [11,12]. The status of PM adsorbed on the leaves could reach the saturation point and block the stoma during the dry winter season [13]. The deposition rate is also sensitive to weather conditions [3,14]. Leaf PM deposition velocity was found to increase with increasing wind speed [15,16], and PM removal rates from the leaf surface were found to be correlated with high rainfall intensity and duration [17]. Therefore, urban green areas have different effects of PM reduction that depend on their structure and the meteorological factors that affect PM movement. It is important to identify the characteristics of PM reduction in urban green areas according to environmental conditions and meteorological factors.

Most studies have focused on identifying the reduction of PM concentration with equations and chamber-based experiments [15–17], so the effects of reducing air pollutants through urban green space need to be verified in the field.

To identify the characteristics of PM reduction in urban green areas, we conducted a study in the buffer green area of the Sihwa Industrial Complex. The research objectives were to determine the effect of urban green area in reducing PM; analyze the relationship between PM, the meteorological factors, and air pollutants; and identify the seasonal patterns of PM reduction by urban green areas.

2. Materials and Methods

2.1. Study Site

The study site, Sihwa (37°22′N, 126°46′E), is a national industrial complex adjacent to the west coast of Korea and consists of steel-, chemical-, and machinery-oriented industries. The main air pollutants of the Sihwa Industrial Complex are NO_2 and SO_2 emissions, which are caused by the operation of vehicles and ships as well as the combustion involved in manufacturing activities during the production process [18]. The Sihwa Industrial Complex has the characteristic of wind blowing in the northwest direction, from the coast to the land. Thus, the buffer green area was established in the 2000s to prevent the spread of air pollutants and odor from the industrial complex to the residential area. The buffer green area is an urban green area located in the center of Sihwa Industrial Complex and New City. The dimensions of this site are 3.46 km (L), 0.18 to 0.25 km (W), and 10 m (H), and it consists of a Japanese black pine (*Pinus thunbergii*) and Japanese red pine (*Pinus densiflora*) community [19]. The densities of the Japanese black pine (*Pinus thunbergii*) and Japanese red pine (*Pinus densiflora*) are 24.5 trees/100 m^2 and 0.75 trees/100 m^2, respectively. To analyze the PM reduction effect of the urban green area, we selected three PM monitoring points: Somang park in the industrial complex (IC; 37°20′ N, 126°43′ E), the urban green area (UG; 37°20′ N, 126°43′ E), and Jungang park in the residential area (RA; 37°20′ N, 126°44′ E) (Figure 1). The distance between IC and UG is 1 km, and the distance between RA and UG is 0.76 km. As IC is surrounded by various industries and RA is a nearby apartment complex, we considered that these measuring points could reflect the characteristics of PM in each area. IC is dominated by *Pinus strobus* (4.5 trees/100 m^2) and *Pseudocydonia sinensis* (0.5 trees/100 m^2). RA is made up of *Pinus rigida* (8.75 trees/100 m^2) and *Pinus densiflora* (0.5 trees/100 m^2).

2.2. Measurement Method

The PM_{10} and $PM_{2.5}$ concentrations were measured by a mobile PM measuring device, Dustmate (Turnkey, UK, ±5% accuracy), which was installed 1.5 m above the ground. The use of Dustmate for analyzing PM concentrations proved to be successful in studies [20,21]. Dustmate is a real-time PM mass concentration meter that uses the light scattering method. The Sihwa Industrial Complex is a coastal area and is highly influenced by onshore winds. Thus, Dustmate is suitable for checking varied PM concentrations in real time and can measure at high concentrations. However, in the case of the light scattering method, the concentration can be overestimated because the material can cause interference in light transmission, which could be highly influenced by humidity [22]. We excluded the PM concentration data with humidity values over 80% to avoid overestimation of PM concentrations due to binding effects of vapor particles and particulate matter. PM monitoring was conducted from April to October 2019; citizens in Korea usually visit urban green areas and enjoy the outdoors during this season. We measured the PM concentration in the morning rush hour (7–9 h), around noontime (11–13 h), and during evening rush hour (16–18 h), considering the peak hours with a large transient population. We measured data at intervals of 1 s and used these data as averages of 5 min. Based on these PM concentration data, we calculated the time mean, monthly average, and reduction rates of PM as follows:

$$\text{PM reduction rate } (\%) = 100 \times \frac{C_{IC} - C_{RA}}{C_{IC}} \quad (1)$$

where C_{IC} is PM concentration in IC and C_{RA} is PM concentration in RA.

Figure 1. The location of measuring points and national atmospheric environmental research station in Siheung City, Gyeonggi-do, Republic of Korea (modified from map.kakao.com). IC: industrial complex; UG: urban green area; RA: residential area; Airkorea-IC: national atmospheric environmental research station located near IC; Airkorea-RA: national atmospheric environmental research station located near RA.

In order to identify the seasonal patterns of PM reduction by UG, the relationship between PM, meteorological factors, and air pollutants was analyzed. We used a portable weather meter to measure temperature, wind direction, wind speed, and humidity at each point for 24 h. Since the Dustmate equipment could not measure the concentrations of air pollutants (ozone (O_3), NO_2, SO_2, carbon monoxide (CO)), these concentrations were measured by the national atmospheric environmental research stations located nearby in IC (Airkorea-IC; 37°20′N, 126°43′E) and RA (Airkorea-RA; 37°20′N, 126°44′E) (Figure 1). We hypothesized that the values from those stations could represent the air pollutant concentrations for the measuring points.

2.3. Analysis Method

We grouped the PM concentrations, meteorological factors, and air pollutants measured by season, i.e., spring (April and May), summer (June and August), and fall (September and October). To analyze the seasonal characteristics of PM reduction in UG, we performed the Pearson correlation analysis among these grouped data by using R v.3.0.2 (R Core Development Team 2019). Statistical analysis using the Tukey HSD procedure of R v.3.0.2 was also conducted on PM concentration and PM reduction rate. The least squares mean was used to test significant differences among measuring points and over month and hour at the 5% probability level.

3. Results

3.1. Monthly PM Concentration and PM Reduction Rate

We analyzed the monthly PM concentration, excluding PM data in July when the humidity exceeded 80%. Regardless of PM measuring points, the highest PM_{10} and $PM_{2.5}$ concentrations were observed in May, with values of 99.7 μg/m^3 and 75.9 μg/m^3, respectively (Figure 2a). During the measuring period, the average PM_{10} and $PM_{2.5}$ concentrations were shown to be at bad levels (daily mean PM_{10}, 80–150 μg/m^3; daily mean $PM_{2.5}$, 35–75 μg/m^3) in May, while they were observed to be at normal levels (daily mean PM_{10}, 30–80 μg/m^3; daily mean $PM_{2.5}$, 15–35 μg/m^3) in April. Thus, the difference between April and May values was high, even though these measurements were collected in the same season. After May, PM levels gradually decreased, with the lowest concentration being observed in August. In September, PM levels began to increase again. Large deviations of PM_{10} and $PM_{2.5}$ concentrations were observed in the spring, meaning that high PM concentration occurred frequently during that time. On the other hand, summer and fall showed a low deviation of PM concentration, indicating that PM levels were maintained relatively constant compared to spring. Furthermore, no significant differences in PM concentration levels were shown in April, May, and June between the measuring points (Figure 2a). Except for these months, high PM levels were shown, following the order of IC, UG, and RA. In August, monthly PM_{10} and $PM_{2.5}$ reduction rates were high, with values of 44.3% and 56.4%, respectively. The monthly PM_{10} and $PM_{2.5}$ reduction rates in October were high as well, reaching 43.5% and 54.2%, respectively. The lowest PM reduction rates were observed in June (Figure 2b).

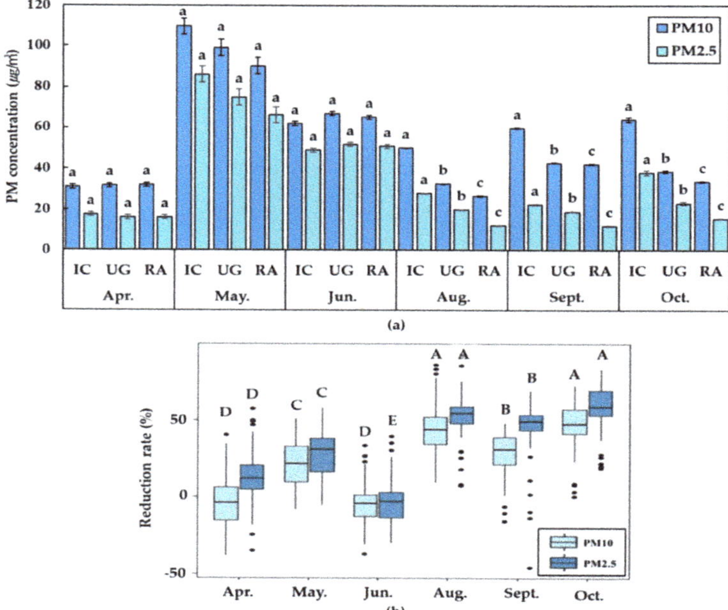

Figure 2. (**a**) Monthly PM concentration (μg/m^3) at measuring points. (**b**) Monthly PM reduction rate (%). Different lowercase letters indicate significant difference between measuring points at a 5% level. Different uppercase letters indicate significant difference between months at a 5% level. IC: industrial complex; UG: urban green area; RA: residential area.

3.2. Hourly PM Concentration and PM Reduction Rate

In the case of hourly PM concentration, the PM_{10} and $PM_{2.5}$ concentrations were the highest during the morning rush hour, reaching 77.9 and 51.4 μg/m^3, respectively (Figure 3a,b). The high PM concentration in the morning rush hour gradually decreased after noon, and the value in the evening rush hour was 30% lower than that in the morning rush hour. PM_{10} and $PM_{2.5}$ reduction rates were the highest (31.4% and 40.3%, respectively) during the evening rush hour, while those of the morning rush hour were the lowest (15.0% and 26.0%, respectively) (Figure 3c).

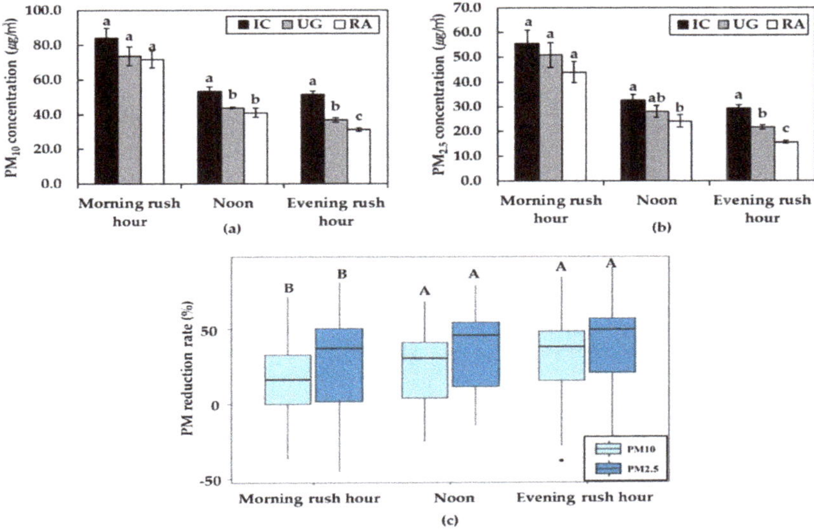

Figure 3. (**a**) Hourly PM_{10} concentration (μg/m^3) at measuring points. (**b**) Hourly $PM_{2.5}$ concentration (μg/m^3) at measuring points. (**c**) Hourly PM reduction rate (%). Different lowercase letters indicate significant difference between measuring points at a 5% level. Different uppercase letters indicate significant difference between hours at a 5% level. IC: industrial complex; UG: urban green area; RA: residential area.

3.3. Relation between PM Concentration, Meteorological Factors, and Air Pollutants by Season

The PM concentration levels are affected by meteorological factors and the amount of emitted air pollutants. In this study, we considered the meteorological factors and air pollutants as factors affecting the PM concentration. Table 1 shows the correlation coefficients between the PM, meteorological factors, and air pollutants. In spring, PM levels shows positive correlation with humidity, NO_2, and CO, while negative correlation was shown with wind speed and O_3. In summer, PM levels were positively correlated with the amount of SO_2 and negatively correlated with the temperature. In the case of fall, correlation between the PM and meteorological factors was the same as in spring, and PM was positively correlated with NO_2, CO, and SO_2.

Table 1. Seasonal correlation coefficients between PM concentration, meteorological factors, and other air pollutants.

Correlation Coefficient	PM_{10}	$PM_{2.5}$	PM_{10}	$PM_{2.5}$	PM_{10}	$PM_{2.5}$
	Spring (April, May)		Summer (June, July)		Fall (September, October)	
Temperature	−0.62	−0.60	−0.86 **	−0.89 **	−0.58	−0.40
Wind speed	−0.95 ***	−0.94 ***	0.54	0.58	−0.82 **	−0.69 *
Humidity	0.96 ***	0.96 ***	0.57	0.57	0.67 *	0.51
O_3	−0.93 ***	−0.91 ***	−0.48	−0.48	−0.80 **	−0.66
NO_2	0.90 **	0.88 **	0.50	0.49	0.96 ***	0.88 **
CO	0.86 **	0.84 **	0.67	0.68 *	0.93 ***	0.87 **
SO_2	0.40	0.36	0.91 ***	0.93 ***	0.38	0.50

Note: * $p < 0.05$, ** $p < 0.01$, *** $p < 0.001$.

4. Discussion

4.1. Factors Affecting the Monthly PM Reduction in Urban Green Areas

Overall, the PM values were higher at IC than at UG and RA. This result could be attributed to the PM reduction effects of UG. As PM generated from IC passed through UG, trees could absorb PM via leaf stomata [8,23] and remove PM by deposition onto the leaves and branches [24,25]. However, we also considered the spatial distance effect on PM reduction according to the distance from pollutant emission areas. PM_{10} and $PM_{2.5}$, which have a long atmospheric lifetime [26], decreased, so we can speculate that the influence of UG on PM reduction was greater than that of distance.

The difference between the PM reduction rates of IC and RA was especially high in August and October. The high PM reduction rate in August seemed to be related not only to PM reduction mechanism of UG but also to an increase of atmospheric mixing height during the summer [27–29]. The dispersion of PM occurred rapidly due to active mixing by air turbulence in the summer [28], leading to a minimum PM concentration. Therefore, it is likely that PM could quickly spread due to the effects of intensive air mixing and prevailing western wind systems, resulting in increasing PM reduction through UG. High PM reduction rates in October were also related to the active atmospheric dispersion affecting the PM reduction, similar to the results observed in August. Moreover, the cleaning effect due to high precipitation in August during the monsoon rainy season should be considered. However, we excluded the effect of PM reduction caused by precipitation because we sampled PM in the dry period, not on rainy days. June, the sharing the summer season with August, had the lowest PM reduction rate. This phenomenon appears to be related to high humidity (remaining in the range of 60% to 70%) in June, whereas other months had lower humidity. The increase of humidity in the atmosphere could reduce solar radiation reaching the earth surface, resulting in adjusting the atmospheric temperature to be similar to that of the earth. As the atmosphere nearer to the surface becomes colder than the upper layers, the air currents seem to decrease and air pollutants become stagnant [30,31]. Therefore, the high humidity interferes with upward-moving air currents, leading to the increase of air pollutant concentrations.

4.2. Factors Affecting the Hourly PM Reduction in Urban Green Areas

High PM levels during the morning rush hour seemed to be related to increased traffic loads. Thus, this result is likely because of the effects of PM generated by vehicles and atmospheric congestion during the early morning. We observed that the PM concentrations in the morning rush hour slightly decreased after noon. This was related to active atmospheric dispersion, as temperature and wind speed slowly increased and humidity decreased after noon. A similar pattern was reported in a metropolitan city in Korea, where high PM concentration and size distribution of PM were affected by increased traffic volumes during the morning rush hour (6–9 h) and then decreased when the wind speed increased [22].

A great reduction in PM concentration during the evening rush hour was related to the strong west wind blowing from IC to RA. It is likely that the PM-reducing effect of UG is greatly increased when PM in IC passes through the UG due to the strong west wind and active atmospheric diffusion after noon. On the other hand, a low PM reduction rate was shown in the morning rush hour in study areas, which was related to the high PM concentration at RA. In the case of RA, with an increase in traffic during the morning rush hour, PM congestion was caused by the east wind blowing and the residential area being surrounded by high apartments [22,32,33]. PM concentration in RA was more affected by high traffic volume and PM congestion than PM from IC during the morning rush hour. The atmospheric diffusion models indicated that air pollutants would not spread actively in the residential area due to the west wind from the coast and the east wind from land coexisting in the study area [34].

4.3. Factors Affecting the PM Concentration by Season

This study showed a high concentration of PM in May. Recently simulated PM concentrations showed a strong negative correlation with regional wind speed, implying that reduced regional ventilation is likely associated with more stagnant conditions that cause severe pollutant episodes in South Korea [35]. Furthermore, temperature and wind speed had a negative correlation with PM, which is related to active atmospheric diffusion with increasing temperature and wind speed [36], and the high temperature influences convective air currents, resulting in rapid dispersion of PM in the atmosphere [29]. It is known that humidity contributes to the increase in PM concentration by acting as the condensation nucleus of air pollutants [37].

PM concentration was positively correlated with NO_2, CO, and SO_2. This is related to the PM formation, as air pollutants are precursors of PM [38,39]. PM concentration in summer had especially high correlation with SO_2, while NO_2 and CO were highly correlated with PM concentration in spring and fall. In summer, as the atmospheric diffusion becomes more active, the SO_2 diffusion produced by the operation of the industrial complex could affect the overall PM concentrations at the measuring points [3,40]. In spring and fall, as NO_2 and CO produced from the vehicles spread in all directions around the roadside, they could have a greater impact on PM concentration when compared to SO_2 from the industrial complex [3,34]. Thus, the analysis of the PM composition is needed to identify the exact path of air pollutants.

O_3 only showed a negative correlation with PM, which might be related to the organic PM generation process through oxidation of volatile organic compounds (VOCs) by O_3 [41]. However, Bell et al. [42] observed a positive correlation between PM and O_3, as PM and O_3 are produced by photochemical reactions. These contradictory results indicate that PM and O_3 levels are driven by a combination of the chemical reactions of their precursors and meteorological factors. Therefore, it is necessary to analyze the correlation between PM and O_3 by identifying the environmental factors affecting these occurrences. In addition, measuring O_3 in urban green areas and analyzing PM formation and extinction data should be conducted to understand the real correlation between O_3 and PM in urban green areas.

5. Conclusions

This study investigated the PM-reducing effect of an urban green area in the Sihwa Industrial Complex. Monthly PM reduction was related to the atmospheric mixing height and west wind blowing. Hourly PM reduction was affected by the speed and direction of wind and the vehicle traffic volumes. Thus, the reduction of PM by the urban green area was associated with PM absorption, adsorption and deposition on leaves, monthly mixing height, and the large scale of air current movement affecting air stagnation and PM dispersion. Our results suggest that an urban green area will reduce PM concentration if the forest layer of the urban green area is planted at a proper location, with changes made as necessary, and the operation of industries and vehicles is managed. However, this study only identified the PM-reducing effect of an urban green area based in a limited space, namely Sihwa Industrial Complex, and analyzed the measured data just for a short period of time. Furthermore, we did not measure PM without an urban green area to confirm the real PM reduction effect of the urban green area. Further research on PM measurement at a control site is required, and it is necessary to identify the overall seasonal PM reduction characteristics through long-term monitoring. Moreover, as the wind condition significantly influences the urban green area's PM reduction effect, further research of prevailing wind field modeling at spatial boundaries on a city level should be carried out to address local solutions for reducing PM concentration.

Author Contributions: Collecting the field data and writing the paper, S.-Y.Y.; collecting the field data and analyzing the statistical data on meteorological factors and pollutants, T.K. and S.H.; managing the research project and contributing to the discussion of wind fields and urban green areas, S.C.; designing the study site and measuring equipment and statistical analyses of all data, C.-R.P. All authors have read and agreed to the published version of the manuscript.

Funding: This research received no external funding.

Acknowledgments: This study was funded by National Institute of Forest Science of Korea, grant number NIFOS FE0000201801.

Conflicts of Interest: The authors declare no conflict of interest.

References

1. Chen, W.Y.; Jim, C.Y. *Assessment and Valuation of the Ecosystem Services Provided by Urban Forests. Ecology, Planning, and Management of Urban Forests*; Springer: New York, NY, USA, 2008.
2. Irga, P.J.; Burchett, M.D.; Torpy, F.R. Does urban forestry have a quantitative effect on ambient air quality in an urban environment? *Atmos. Environ.* **2015**, *120*, 173–181. [CrossRef]
3. Song, Y.-B. Effect of Green Buffer Zone in Reducing Gaseous Air Pollutants in the Shiwha Industrial Area. *J. Korean Inst. Landsc. Archit.* **2006**, *33*, 90–97.
4. Vos, P.E.J.; Maiheu, B.; Vankerkom, J.; Janssen, S. Improving local air quality in cities: To tree or not to tree? *Environ. Pollut.* **2013**, *183*, 113–122. [CrossRef] [PubMed]
5. Hwang, K.-I.; Han, B.-H.; Kwark, J.-I.; Park, S.-C. A Study on Decreasing Effects of Ultra-fine Particles (PM2.5) by Structures in a Roadside Buffer Green—A Buffer Green in Songpa-gu, Seoul. *J. Korean Inst. Landsc. Archit.* **2018**, *46*, 61–75. [CrossRef]
6. Zheng, D.; Ducey, M.J.; Heath, L.S. Assessing net carbon sequestration on urban and community forests of northern New England, USA. *Urban For. Urban Green.* **2013**, *12*, 61–68. [CrossRef]
7. Choi, T.-Y.; Moon, H.-G.; Kang, D.-I.; Cha, J.-G. Analysis of the Seasonal Concentration Differences of Particulate Matter According to Land Cover of Seoul - Focusing on Forest and Urbanized Area -. *J. Environ. Impact Assess* **2018**, *27*, 635–646.
8. Janhäll, S. Review on urban vegetation and particle air pollution - Deposition and dispersion. *Atmos. Environ.* **2015**, *105*, 130–137. [CrossRef]
9. Nowak, D.J.; Crane, D.E.; Stevens, J.C. Air pollution removal by urban trees and shrubs in the United States. *Urban For. Urban Green.* **2006**, *4*, 115–123. [CrossRef]
10. Salmond, J.A.; Williams, D.E.; Laing, G.; Kingham, S.; Dirks, K.; Longley, I.; Henshaw, G.S. The influence of vegetation on the horizontal and vertical distribution of pollutants in a street canyon. *Sci. Total Environ.* **2013**, *443*, 287–298. [CrossRef]
11. Pataki, D.E.; Carreiro, M.M.; Cherrier, J.; Grulke, N.E.; Jennings, V.; Pincetl, S.; Pouyat, R.V.; Whitlow, T.H.; Zipperer, W.C. Coupling biogeochemical cycles in urban environments: Ecosystem services, green solutions, and misconceptions. *Front. Ecol. Environ.* **2011**, *9*, 27–36. [CrossRef]
12. Hong, S.; Kang, R.; An, M.; Kim, J.; Jung, E. Study on the Impact of Roadside Forests on Particulate Matter between Road and Public Openspace in front of Building Site - Case of Openspace of Busan City hall in Korea -. *Korean J. Environ. Ecol.* **2018**, *32*, 323–331. [CrossRef]
13. Shackleton, K.; Bell, N.; Smith, H.; Davies, L. *The Role of Shrubs and Perennials in the Capture and Mitigation of Particulate Air Pollution in London*; Centre for Environmental Policy: London, UK, 2010.
14. Kim, S. Analytic Model for Concentration Deficit Profile Caused by a Large Vegetated Area. *J. Korean Soc. Atmos. Environ.* **2000**, *16*, 539–544.

15. Nowak, D.J.; Hirabayashi, S.; Bodine, A.; Hoehn, R. Modeled PM 2. 5 removal by trees in ten U. S. cities and associated health effects. *Environ. Pollut.* **2013**, *178*, 395–402. [CrossRef] [PubMed]
16. Han, D.; Shen, H.; Duan, W.; Chen, L. A review on particulate matter removal capacity by urban forests at different scales. *Urban For. Urban Green.* **2020**, *48*, 126565. [CrossRef]
17. Zhang, L.; Zhang, Z.; Chen, L.; McNulty, S. An investigation on the leaf accumulation-removal efficiency of atmospheric particulate matter for five urban plant species under different rainfall regimes. *Atmos. Environ.* **2019**, *208*, 123–132. [CrossRef]
18. Korea Environment Institute. Calculation of PM2.5 Contribution Concentrations and Early Deaths due to Operation of Major National Industrial Complexes. *Nihon Shoni Arerugi Gakkaishi. Japanese J. Pediatr. Allergy Clin. Immunol.* **2018**, *22*, 1–15. [CrossRef]
19. Choi, J.-W. A Study on Vegetation Changes for 11years and Vegetation Structure in the Green Buffer Zone of Sihwa Industrial Complex. *J. Korean Soc. Environ. Restor. Technol.* **2018**, *21*, 81–96. [CrossRef]
20. Chen, J.; Yu, X.; Sun, F.; Lun, X.; Fu, Y.; Jia, G.; Zhang, Z.; Liu, X.; Mo, L.; Bi, H. The concentrations and reduction of airborne particulate matter (PM10, PM2.5, PM1) at shelterbelt site in Beijing. *Atmosphere (Basel).* **2015**, *6*, 650–676. [CrossRef]
21. Wu, Y.; Liu, J.; Zhai, J.; Cong, L.; Wang, Y.; Ma, W.; Zhang, Z.; Li, C. Comparison of dry and wet deposition of particulate matter in near-surface waters during summer. *PLoS ONE* **2018**, *13*, 1–15. [CrossRef]
22. Cho, B.-Y.; Shin, S.-H.; Jung, C.-S.; Ju, M.-H.; Yoon, M.-H.; Ahn, J.-E.; Bae, G.-S. Characteristics of Particle Size Distribution at the Roadside of Daegu. *J. Korean Soc. Atmos. Environ.* **2019**, *35*, 16–26. [CrossRef]
23. Hofman, J.; Bartholomeus, H.; Janssen, S.; Calders, K.; Wuyts, K.; Van Wittenberghe, S.; Samson, R. Influence of tree crown characteristics on the local PM10 distribution inside an urban street canyon in Antwerp (Belgium): A model and experimental approach. *Urban For. Urban Green.* **2016**, *20*, 265–276. [CrossRef]
24. Province, G.; Qiu, Y.; Guan, D.; Song, W.; Huang, K. Chemosphere Capture of heavy metals and sulfur by foliar dust in urban Huizhou. *Chemosphere* **2009**, *75*, 447–452. [CrossRef]
25. Ottosen, T.B.; Kumar, P. The influence of the vegetation cycle on the mitigation of air pollution by a deciduous roadside hedge. *Sustain. Cities Soc.* **2020**, *53*, 101919. [CrossRef]
26. Gugamsetty, B.; Wei, H.; Liu, C.N.; Awasthi, A.; Tsai, C.J.; Roam, G.D.; Wu, Y.C.; Chen, C.F. Source Characterization and Apportionment of PM10, PM2.5 and PM0.1 by Using Positive Matrix Factorization. *Aerosol Air Qual. Res.* **2012**, *12*, 476–491. [CrossRef]
27. Väkevä, M.; Hämeri, K.; Puhakka, T.; Nilsson, E.D.; Hohti, H.; Mäkelä, J.M. Effects of meteorological processes on aerosol particle size distribution in an urban background area. *J. Geophys. Res. Atmos.* **2000**, *105*, 9807–9821. [CrossRef]
28. Hu, M.; Liu, S.; Wu, Z.-J.; Zhang, J.; Zhao, Y.-L.; Wehner, B.; Wiedensolher, A. Effects of high temperature, high relative humidity and rain process on particle size distributions in the summer of Beijing. *Huan Jing Ke Xue Huanjing Kexue* **2006**, *27*, 2293–2298. [CrossRef]
29. Li, Y.; Chen, Q.; Zhao, H.; Wang, L.; Tao, R. Variations in pm10, pm2.5 and pm1.0 in an urban area of the sichuan basin and their relation to meteorological factors. *Atmosphere (Basel).* **2015**, *6*, 150–163. [CrossRef]
30. Bathmanabhan, S.; Saragur Madanayak, S.N. Analysis and interpretation of particulate matter-PM10, PM2.5 and PM1 emissions from the heterogeneous traffic near an urban roadway. *Atmos. Pollut. Res.* **2010**, *1*, 184–194. [CrossRef]
31. Jayamurugan, R.; Kumaravel, B.; Palanivelraja, S.; Chockalingam, M.P. Influence of Temperature, Relative Humidity and Seasonal Variability on Ambient Air Quality in a Coastal Urban Area. *Int. J. Atmos. Sci.* **2013**, *2013*, 1–7. [CrossRef]
32. Laakso, L.; Hussein, T.; Aarnio, P.; Komppula, M.; Hiltunen, V.; Viisanen, Y.; Kulmala, M. Diurnal and annual characteristics of particle mass and number concentrations in urban, rural and Arctic environments in Finland. *Atmos. Environ.* **2003**, *37*, 2629–2641. [CrossRef]
33. Kim, J.-A.; Jin, H.-A.; Kim, C.-H. Charateristics of Time Variations of PM10 Concentrations in Busan and Interpreting Its Generation Mechanism Using Meteorological Variables. *J. Environ. Sci.* **2007**, *16*, 1157–1167.
34. Ko, J.-C.; Ryu, K.-W.; Park, I.G. The Dispersion Characteristics Analysis of PM10 and PM2.5 In Industrial Complex using Atmospheric Dispersion Model. *J. Korean Soc. Environ. Technol.* **2018**, *19*, 366–374. [CrossRef]
35. Kim, H.C.; Kim, S.; Kim, B.; Jin, C.; Hong, S.; Park, R.; Son, S.; Bae, C.; Bae, M.; Song, C.; et al. Recent increase of surface particulate matter concentrations in the Seoul Metropolitan Area, Korea. *Sci. Rep.* **2017**, 1–7. [CrossRef] [PubMed]

36. Lin, J.; Liu, W.; Li, Y.; Bao, L.; Li, Y.; Wang, G.; Wu, W. Relationship between meteorological conditions and particle size distribution of atmospheric aerosols. *J. Meteorol. Environ.* **2009**, *25*, 1–5.
37. Shin, M.-K.; Lee, C.-D.; Ha, H.-S.; Choe, C.-S.; Kim, Y.-H. The Influence of Meteorological Factors on PM10 Concentration in Incheon. *J. Korean Soc. Atmos. Environ.* **2007**, *23*, 322–331.
38. Seinfeld, J.H.; Pandis, S.N. *Atmospheric Chemistry and Physics: From Air Pollution to Climate Change*, 3rd ed.; John Wiley & Sons: Hoboken, NJ, USA, 2016.
39. Xu, X.; Kim, J.-O. Planting Design Strategies and Green Space Planning to Mitigate Respirable Particulate Matters - Case Studies in Beijing, China -. *J. Korean Inst. Landsc. Archit.* **2017**, *45*, 40–49.
40. Suh, M.; Song, Y.-B. *A Study on the Influence of Sihwa Industrial Complex and Air Pollution in Sihwa District*; Siheung-si Environmental Technology Development Center: Gyeonggi-do, Korea, 2004.
41. Bae, M.S.; Schauer, J.J.; Lee, T.; Jeong, J.H.; Kim, Y.K.; Ro, C.U.; Song, S.K.; Shon, Z.H. Relationship between reactive oxygen species and water-soluble organic compounds: Time-resolved benzene carboxylic acids measurement in the coastal area during the KORUS-AQ campaign. *Environ. Pollut.* **2017**, *231*, 1–12. [CrossRef]
42. Bell, M.L.; Kim, J.Y.; Dominici, F. Potential confounding of particulate matter on the short-term association between ozone and mortality in multisite time-series studies. *Environ. Health Perspect.* **2007**, *115*, 1591–1595. [CrossRef]

Article

Reconstruction of Resin Collection History of Pine Forests in Korea from Tree-Ring Dating

En-Bi Choi [1], Yo-Jung Kim [2], Jun-Hui Park [1], Chan-Ryul Park [3,*] and Jeong-Wook Seo [1,*]

1 Department of Forest Product, Chungbuk National University, Cheongju 28644, Korea; chenbi0715@chungbuk.ac.kr (E.-B.C.); fairy@chungbuk.ac.kr (J.-H.P.)
2 Department of Wood and Paper Science, Chungbuk National University, Cheongju 28644, Korea; wnsgml9547@chungbuk.ac.kr
3 Urban Forests Research Center, National Institute of Forest Science, Seoul 02455, Korea
* Correspondence: maeulsoop@korea.kr (C.-R.P.); jwseo@chungbuk.ac.kr (J.-W.S.); Tel.: +82-2-961-2612 (C.-R.P.); +82-43-261-2543 (J.-W.S.)

Received: 29 September 2020; Accepted: 30 October 2020; Published: 2 November 2020

Abstract: Resin is one of the traditional non-timber forest products in the Republic of Korea. In order to investigate the chronological activity of resin collection, the wounds/cuts on red pines (*Pinus densiflora*) were dated using a tree-ring analysis technique. Additionally, the size of the trees in the resin collection years and the present conditions of the trees were investigated to verify the tree conditions and the size of wounds. Eighty-eight red pines distributed over nine sites in the Republic of Korea were selected to extract increment cores and investigate the wound size. Through the tree-ring analysis, the trees with big wounds (24.7 × 104.7 cm) made via panel hacksaw method were dated in the range 1938–1952, whereas small wounds (40.2 × 20.9 cm) made via the conventional chisel method were dated between 1956 and 1973. Moreover, the red pines thicker than 20.0 cm were the ones that were used for resin collection. Furthermore, the wounds created by the conventional chisel were healed with time, whereas the ones formed via the panel hacksaw method still required long times for healing. The large wounds had the advantage of supplying a large amount of resin, but this was temporary. On the other hand, the smaller wounds formed via the traditional chisel method could generate resin for a longer time and heal faster.

Keywords: resin collection; red pines; wounds; panel hacksaw method; conventional chisel method

---

1. Introduction

Resin is one of the traditional non-timber forest products (NTFPs) which represents all kinds of non-timber resources produced in forests [1] in the Republic of Korea as in other countries. The use of NTFPs has remained in the form of knowledge delivered from generation to generation in a local community, rendering the community as a fundamental data source of the NTFPs. In the records of traditional forest knowledge (www.koreantk.com), resin is defined as a raw material used in traditional crafts, medicine, foods, life technology, and agriculture [2]. Various uses of pine resin with honey mel can be found in the old literature "*CHISENGYORAM*" (literally means the guideline for management of livelihood, 1691) [3], "*GOSASINSEO*" (literally means the new book for deep thinking, 1771) [4], and "*NONJEONGHOIYO*" (literally means the technical book for agriculture and forestry, 1831) [5] in Korea.

Korea was under Japanese rule between 1910 and 1945, and thus, Korea was greatly influenced by Japan's socioeconomic policy during this period. Japan entered war in the late 1930s and implemented the first seven-year plan to produce synthetic oil in 1937 [6]. The seven-year plan was implemented by the Japanese government because sufficient synthetic oil was not produced by the domestic companies to support fuel for weapons [7]. To obtain resin from red pines (*Pinus densiflora*), the panel hacksaw method, in which a cut of around 150 cm length was made, was applied. On the other hand, in the traditional method in the Republic of Korea, the so-called chisel method was used, whereby a cut was made of approx. 6 cm width and 12–21 cm length [2].

The tree-ring analysis technique, also known as dendrochronology, is a powerful tool to date annual rings in woody plants. In the dating technique, the ring-width time series patterns obtained from the same tree species under similar growing conditions are synchronized with each other. This is called cross-dating in dendrochronology [8,9]. Cross-dating has been broadly applied to date annual rings in forest trees to investigate their dead and/or wounded years [10,11] as well as use their wood in archaeological architecture [12], wood craft [13], picture frames [14], musical instruments [15], and tracing wood trade [16]. The dating results of the cuts in tree rings could provide historical information on natural events, e.g., forest fire [17,18], landslide [19], flash flood [20,21], insect damage [22], and human activities, e.g., resin or latex collections.

The primary aim of the present study was to date resin collection from red pines (*Pinus densiflora*), estimate the tree size in the years when the resin was collected, and verify the current conditions of the trees. The results would serve as a reference to understand the changes in resin collection methods by year and determine forest policy on the sustainable usage of forest resources by resin collection methods.

2. Materials and Methods

2.1. Sampling Sites and Trees

To verify records on resin collection in ancient literature and recent public reports, the National Institute of Forest Science in the Republic of Korea conducted a survey for two years (2017–2018) on the distribution and growth conditions of red pine trees (*Pinus densiflora*) which were wounded for collecting resin. The survey revealed that the wounded pines were distributed in 43 regions in the Republic of Korea [2].

In the present work, 9 study sites were selected based on their geographical distribution (Figure 1). Except Anmyeondo (AM) and Haeinsa (HI), only one site was selected in each province. In AM and HI, we selected 2 and 3 sites, respectively, due to the presence of a large number of wounded trees there. All sites were located either in a village or close to a temple, public park, or recreation forest, where there was easy accessibility (Table 1).

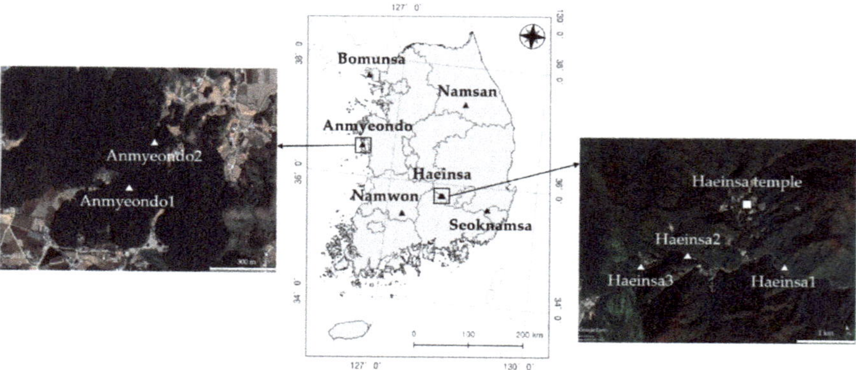

Figure 1. The locations of the sampling sites (triangles).

Table 1. Description of the sampling sites and trees (*Pinus densiflora*).

Sites	ID	Geography and Environment			Sample Trees		
		Latitude	Longitude	Elevation (m a.s.l)	Diameter(cm)		No. (n)
Bomunsa	BM	37°41′ N	126°19′ E	67–88	46.9	(±4.2)	10
Namsan	NS	37°21′ N	128°23′ E	318–334	50.7	(±4.6)	10
Anmyeondo	AM1	36°30′ N	126°21′ E	13–35	61.6	(±6.3)	10
	AM2	36°30′ N	126°21′ E	39–53	52.8	(±6.0)	10
Haeinsa	HI1	35°47′ N	128°06′ E	486–500	62.3	(±8.9)	10
	HI2	35°47′ N	128°05′ E	591–603	50.5	(±8.0)	10
	HI3	35°47′ N	128°04′ E	691–726	51.3	(±5.3)	10
Seoknamsa	SN	35°37′ N	129°02′ E	279–309	62.5	(±11.2)	10
Namwon	NW	35°26′ N	127°19′ E	145–154	60.2	(±13.4)	8

To examine the dating of resin collection and measure the wound size, 10 trees were studied at each site, except Namwon (NW), where 8 trees were used (Table 1). Mean diameters larger than 60 cm were obtained in 4 sites (AM1, HI1, SN, and NW), between 50 and 60 cm in 4 sites (NS, AM2, HI2, and HI3), and smaller than 50 cm in only one site (BM). The diameters were the means of the diameters from approximately 10 cm above and below the wounds.

2.2. Measuring the Wound Size and Collecting Increment Cores

The maximum height and width of a wound measured was reported as the size of the wound made in red pines (*Pinus densiflora*) for resin collection (Figure 2A). To establish the ring-width time series for a wounded surface, an increment core was extracted from the outermost surface of a wound (core A in Figure 2B). Likewise, to establish a reference time series to date the ring-width time series of core A, an increment core was also extracted from the opposite side, which contained tree rings continually from the current year to the year when growth started (core B in Figure 2B).

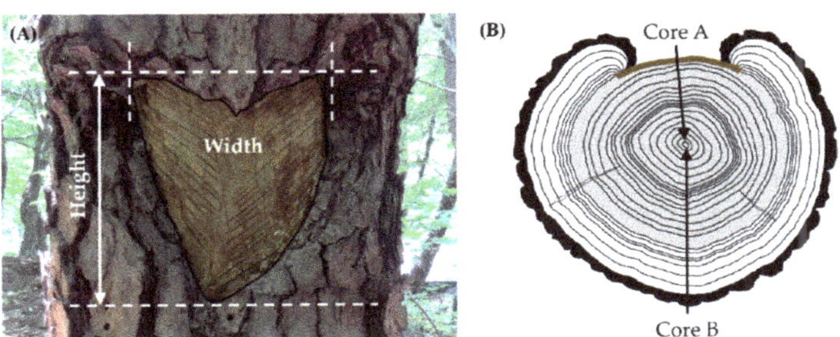

Figure 2. (**A**) Measuring the size of the wound for resin collection and (**B**) extraction of the increment core from a wounded surface (core A) and a sound surface (core B).

2.3. Sample Prepartion and Ring-Width Measurement

Before mounting, the collected increment cores were dried in air to avoid their shrinkage in the mount. When the dried cores were mounted on wooden sticks, the direction of the tracheids was kept vertical. The cross plans of the mounted cores were sanded using a belt sander unless the ring boundaries were clearly visible. The sanding was started with #80 through #120 and #360 up to #600. The ring boundaries were observed under a stereo microscope (Nikon SMZ, Japan), and the annual ring width was measured to the nearest 0.01 mm using the LINTAB (Rinntech, Germany) measurement system.

2.4. Cross-Dating

Cross-dating is a technique used to date tree rings [23] by testing the synchronization between the individual ring-width time series. In order to verify the synchronization using statistical models, t-value [24] and G-value [25] in the TSAP-Win program (Rinntch, Germany) were applied. The t-value (Equation (1)) was developed based on the number of overlapped years between the individual ring-width time series and their correlation coefficient, while the G-value (Equation (2)) was based on the year-to-year agreement between them.

$$t = \frac{r \times \sqrt{n-2}}{\sqrt{(1-r^2)}} \tag{1}$$

where r is the correlation coefficient between the individual ring-width time series and n is the number of overlapped years.

$$G_{(x,y)} = \frac{1}{n-1} \sum_{i=1}^{n-1} \left[G_{ix} + G_{iy} \right] \tag{2}$$

If $(x_{i+1} - x_i) > 0$, $G_{ix} = +1/2$, $(y_{i+1} - y_i) > 0$, $G_{iy} = +1/2$,
$(x_{i+1} - x_i) = 0$, $G_{ix} = 0$, $(y_{i+1} - y_i) = 0$, $G_{iy} = 0$,
$(x_{i+1} - x_i) < 0$, $G_{ix} = -1/2$, $(y_{i+1} - y_i) < 0$, $G_{iy} = -1/2$

where $G_{(x,y)}$ is the G-value and x_i and y_i are the measurement ring-width values for the ith year.

The cross-dating was considered successful when the t- and/or G-values were higher than 3.5 and 65%, respectively; however, the final decision was made by synchronization between the time series through human eyes.

The local master chronologies to date the resin collection were developed using the increment cores from the opposite sides of the resin collection. The year of resin collection was assigned based on the t- and G-values between the individual ring-width time series and the corresponding local master chronologies.

2.5. Resin Collection Season

To determine the resin collection season, the phases of cell development in the outmost tree rings were observed under stereoscopic microscope. In the Republic of Korea, red pines growing at low altitude begin to form the annual ring in March and end between October and November [26,27]. In the temperate zone, the latewood formation in conifer trees begins in June/July [28]. Based on the past reports, the resin collection season was determined as follows: for those where only earlywood was observed in the outermost tree rings (Figure 3A), the resin collection was done between spring and early summer; for those where incomplete latewood formation was observed (Figure 3B), between late summer and autumn; and for those where complete latewood formation was observed (Figure 3C), between autumn of the current year and spring of the next year.

Figure 3. Stereoscopic observation of the outermost annual ring. (**A**) Earlywood (EW); (**B**) uncompleted latewood (UCLW); (**C**) completed latewood (CLW); PTR: previous annual ring.

2.6. Diameter of the Red Pines up to the Resin Collection Year

The diameter of the red pines during resin collection years was estimated using the increment cores extracted from the wounded side and the opposite side. The former cores were used to measure the half diameter from the pith to the wounded surface (a or a′ in Figure 4B) and the latter ones were from the pith to the tree rings formed up to the resin collection year (b or b′ in Figure 4). When the cores from the wounded side and/or the other side had pith, the half diameter up to the resin collection year was measured using the lengths of a and/or b. By contrast, when both or one core had no pith, the half diameter was measured as a′ and/or b′. The pith location to obtain a′ and b′ was estimated based on the arc of the innermost tree ring (dark black arc in B of Figure 4). The current-year thickness of the bark was applied to the bark thickness at the resin collection year. Therefore, the diameter up to the resin collection year was estimated using Equation (3).

$$D\ (cm) = a + b + 2c \text{ or } D\ (cm) = a' + b' + 2c \quad (3)$$

where D is the diameter at the resin collection year, a or a' is the observed or estimated length from the pith to the outermost tree ring of the wounded side, b or b' is the observed or estimated length from the pith to the tree ring formed at the resin collection year at the opposite side of the wound, and c is the bark thickness at the current year.

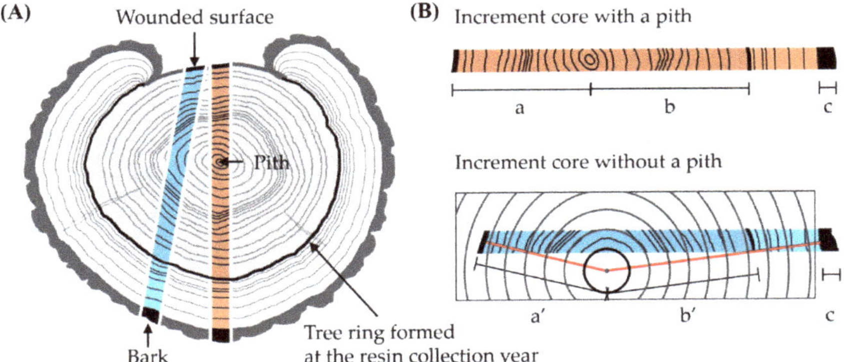

Figure 4. Illustration to estimate the diameter of red pines (*Pinus densiflora*) from the wounds of resin collection. (**A**) Cross-section with directions to extract increment cores; (**B**) estimation of the diameter at resin collection year for cores with or without pith.

The estimated dimeter (D in Figure 4), which lay extremely outside the overall distribution, had been removed from further analysis since an outlier can overestimate or underestimate the result. An outlier is determined as follows.

$$Outlier < Q_1 - 1.5 \times (Q_3 - Q_1) \ or > \ Q_3 + 1.5 \times (Q_3 - Q_1) \quad (4)$$

where Q_1 is the first quartile diameter and Q_3 is the third quartile diameter.

3. Results

3.1. Size of the Wounds

The wound size was categorized into two groups. Group 1, wounded by the panel hacksaw method (G1), had wounds higher and narrower than 90 and 40 cm in height and width, respectively, whereas Group 2, wounded by the traditional chisel method (G2), comprised wounds which were lower and wider than 90 and 40 cm in length and width, respectively (Table 2). Although HI3 had wounds narrower than 40 cm in width, it was categorized into G2 based on the height, which was the lowest height among all sites.

Table 2. The mean height and width with standard deviations, the mean areas (brown color) with standard deviations (dotted line), calculated from the height and width (unit: cm).

Sites		BM	NS	AM1	AM2	HI1	HI2	HI3	SN	NW
Group		G1	G1	G2	G 2	G 2	G 2	G 2	G 1	G 1
Wounds	Height	94.0 ±24.0	113.4 ±35.2	32.1 ±9.5	28.5 ±6.3	22.1 ±14.4	12.1 ±5.2	9.8 ±8.4	100.6 ±18.5	110.6 ±21.1
	Width	22.6 ±6.9	19.5 ±5.3	42.8 ±10.8	44.1 ±7.5	43.6 ±17.3	44.8 ±18.6	25.7 ±23.4	21.5 ±8.1	35.0 ±10.4
	Areas									

BM: Bomunsa, NS: Namsan, SN: Seoknamsa, HI: Haeinsa, NW: Namwon, AM: Anmyeondo.

The mean height and width of G1 were 104.7 ± 9.0 cm and 24.7 ± 7.0 cm, respectively, whereas the mean height and width of G2 were 20.9 ± 9.8 cm and 40.2 ± 8.1 cm, respectively. Therefore, G1 had approximately 5 times higher wound height than G2, whereas G2 had approximately 1.6 times wider width than G1. The largest wound area calculated by the highest and widest values at each wound was obtained for NW, followed by NS, SN, BM, AM1, AM2, HI1, HI2, and HI3.

3.2. Dating the Resin Collection from the Trees

Based on the t- and G-values between the individual ring-width time series and the corresponding local master chronologies, each annual ring was given an exact calendar year (Table 3). Through the statistical tests and synchronization test between the ring-width time series from the wounds and the opposite sides and/or the corresponding local master chronologies (Table 3 and Figure 5), the years of resin collection were successfully dated for 83 red pines out of a total 88 trees (Table 4). Among the successfully dated trees, five trees in NS were dated by comparing their ring-width time series with the local master chronology because their time series were not enough long for t- and G-tests. Finally, from the wood cell development phases from the wounds to the outermost annual ring, the resin collection seasons were successfully determined (Table 4).

Table 3. Statistical analysis of individual ring-width time series and the corresponding local master chronology ($p < 0.05$).

Sites	T-Value			G-Value		
	Mean	Max	Min	Mean	Max	Min
BM	8.4	12.3	6.0	63.6	70.0	58.0
NS	6.5	7.6	5.3	60.5	62.0	59.0
AM1	14.2	20.6	10.3	77.9	87.0	70.0
AM2	11.4	15.9	4.8	81.9	88.0	78.0
HI1	11.4	20.5	4.4	74.7	78.0	68.0
HI2	15.8	18.5	10.3	73.9	80.0	60.0
HI3	10.9	17.7	6.1	75.3	87.0	66.0
SN	3.8	5.5	2.6	68.5	79.0	62.0
NW	5.7	13.6	3.1	70.9	78.0	65.0

BM: Bomunsa, NS: Namsan, SN: Seoknamsa, HI: Haeinsa, NW: Namwon, AM: Anmyeondo.

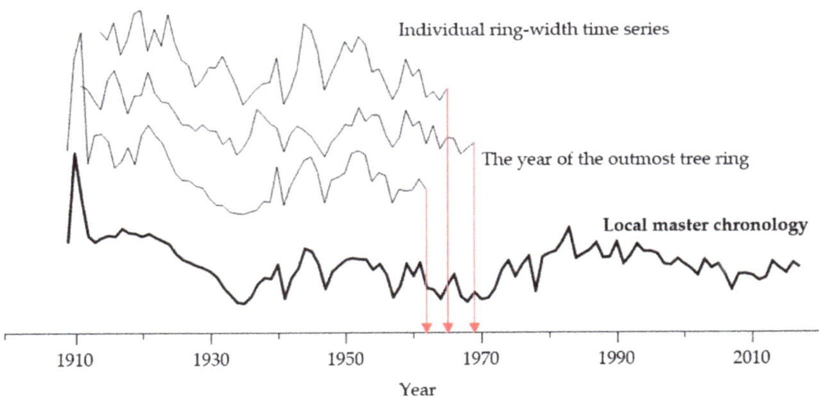

Figure 5. An example of synchronizations between individual ring-width times series from the wound surfaces and the local master chronology (AM1).

Table 4. The dated years and seasons for resin collection.

Sites	Resin Collection Periods				Dated Trees (n)
	Beginning		End		
BM	1944	Spring–Early summer	1952	Fall–Next spring	9
NS	1941	Spring–Early summer	1944	Fall–Next spring	7
AM1	1962	Spring–Early summer	1972	Fall–Next spring	10
AM2	1964	Late summer–Fall	1973	Late summer–Fall	10
HI1	1940	Spring–Early summer	1944	Late summer–Fall	10
HI2	1956	Spring–Early summer	1961	Late summer–Fall	10
HI3	1958	Spring–Early summer	1965	Late summer–Fall	10
SN	1941	Spring–Early summer	1945	Fall–Next spring	10
NW	1938	Fall–Next spring	1942	Fall–Next spring	7

BM: Bomunsa, NS: Namsan, SN: Seoknamsa, HI: Haeinsa, NW: Namwon, AM: Anmyeondo.

All the sites in G1 and HI1 in G2 showed that the resin collection occurred between autumn 1938 and autumn 1944, i.e., almost at the end of the Japanese colonial period (1910–1945). The resin collection in the other sites in G2, except HI1, was between spring 1956 and late summer 1964, i.e., after the Korean War (1950–1953) (Figure 6). Usually, the resin collection in G1 was longer than that in G2, namely in G1 for 5.6 (±1.9) years and in G2 for 8.8 (±2.2) years.

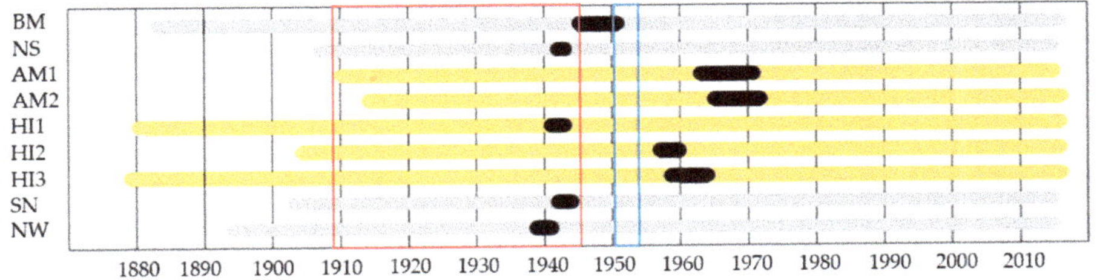

Figure 6. The length of local master chronologies (: G1, : G2) in different resin collection years (: resin collection periods). ☐: the Japanese colonial period, ☐: the Korean War, BM: Bomunsa, NS: Namsan, SN: Seoknamsa, HI: Haeinsa, NW: Namwon, AM: Anmyeondo.

Anatomical investigation revealed that 53.0% red pines (44 out of 83 trees) were subjected to resin collection between spring and early summer, 24.1% (20 trees) between late summer and autumn, and 22.9% (19 trees) between autumn of the current year and spring of the next year.

3.3. Diameter of the Red Pines at Resin Collection

The largest diameter (36.0 cm) at the resin collection year was recorded for NW, followed by HI1 (34.6 cm), AM1 (33.6 cm), HI2 (30.5 cm), SN (29.4 cm), AM2 (28.9 cm), BM (28.1 cm), HI3 (26.0 cm), and NS (25.0 cm) (Table 5). To avoid over- or underestimation due to the outlier, three mean diameter values, viz. AM2 (19.0 cm), SN (18.0 cm), and BM (33.7 cm), were removed from further analysis.

Table 5. Estimated diameters (cm) of the red pines at resin collection years.

Sites	Mean	Max	Min	Sites	Mean	Max	Min
BM	28.1	30.2	25.8	HI1	34.6	44.1	25.6
NS	25.0	26.9	21.9	HI2	30.5	41.7	25.6
AM1	33.6	40.1	21.5	HI3	26.0	35.3	20.5
AM2	28.9	33.4	25.4	SN	29.4	35.7	21.7
				NW	36.0	45.1	25.8

BM: Bomunsa, NS: Namsan, SN: Seoknamsa, HI: Haeinsa, NW: Namwon, AM: Anmyeondo.

4. Discussion

The results of the current study revealed that the size of the wounds for resin collection from red pines could be temporally divided into the Japanese colonial period and post Korean War period. Especially, from the end of the 1930s to the middle of the 1940s, resin collection was very intensive in the Republic of Korea to support the fuel for Japanese weapons [7]. To effectively collect a huge amount of resin, the panel hacksaw method was applied, which makes larger wounds (Figure 7A) than the traditional chisel method in Korea (Figure 7B). This information can be mainly verified from historical documents.

Figure 7. Wounds by (**A**) panel hacksaw method and (**B**) conventional chisel method.

The panel hacksaw method was mainly used between 1938 and 1945. Only BM in G1 (Table 2) showed such hacksaw wounds dated between 1944 and 1952 (Table 5). BM is located in a small island which is not well accessible from the inland. Due to BM's geographical location, the panel hacksaw method has been applied there later than other sites [2]. Unlike the G1 sites, one site in G2 (Table 2), viz. HI1, was dated between 1940 and 1944 for the resin collection. Haeinsa is one of the three Jewel Temples in the Republic of Korea, which was listed as a UNESCO World Heritage Site in 1995. Therefore, from a religious standpoint, the panel hacksaw method which makes large wounds might have not been used in Haeinsa.

The annual rings in the red pines comprised earlywood formed between spring and summer, and latewood formed between summer and autumn, similar to the conifer tree species in the northern hemisphere [26,27]. Therefore, the early- or latewood in the outermost annual ring can give seasonal information on resin collection. However, when the resin collection is determined, the wood cell development process needs to be considered. The development of the wood cells passes through the following processes: cell division in the cambium, cell expansion or elongation, cell wall thickening, cell wall sculpturing, lignification [29]. When wounds are made, the mature wood cells remain intact but the cambial cells, and the cells involved in cell expansion/elongation, cell wall thickening, and lignification, can be easily removed. These cells can be easily destroyed even by a small physical force [30]. Therefore, in most cases, the cells which are physically stable can be observed at the outermost tree ring on the wound surface. Due to this reason, the resin collection seasons can be determined one season earlier.

Most records and research on resin collection focus on the wound size and/or the technique and/or years of resin collection [2,31]. Although the diameters of red pines at the resin collection years can provide information about the forest conditions related to the trees at that time, it has not been well investigated. In the present study, the smallest and largest diameters of the red pines at each site varied from 20.5 (HI3) to 25.8 (BM and NW) cm and from 26.9 (NS) to 45.1 (NW) cm, respectively. These results indicated that the red pines thicker than 20.0 cm were used for resin collection, while trees thicker than 45.0 cm were rare.

Trees have an ability to heal wounds by forming callus tissue around the edges of the wound [32]. The duration to completely cover a wound is strongly related to the size of the wound. The current study found that some wounds created by the conventional chisel method could successfully seal the wounds (Figure 7B); however, the wounds caused by the panel hacksaw method required a long time (Figure 7A). When natural resources are obtained from trees through wounding, the wound size is considered based on the sealing ability of trees.

The panel hacksaw method is useful in increasing the collected amount of pine resin in the short term. However, it left irrevocable wounds which still exist in Korea. Unlike the panel hacksaw method, the conventional chisel method cannot produce a huge amount resin within a short period; however, it offers the advantage of sustainable collection of resin. Sustainable use of non-timber products has countless value and should be transferred to the next generation to maintain a safe and happy life.

5. Conclusions

The current study found that the panel hacksaw method was temporarily applied to collect resin intensively from pine trees under Japanese rule, and after the Korean war, the conventional chisel method was again applied to collect resin until the middle of the 1970s for livelihood. For resin collection, red pines thicker than 20.0 cm were used. The wounds created by the conventional chisel were found to be healed, whereas the ones created via the panel hacksaw method still required long times for healing. From the large wounds, a benefit is that a large amount of resin can be obtained, but these wounds take a long time to heal completely. On the other hand, the traditional chisel method offers a sustainable supply of resin and rapid healing of the wounds. Therefore, we can conclude that the traditional method to obtain non-timber forest products will be a remarkable reference to determine forest policy on the usage of forest resources sustainably and should be passed from generation to generation in Korea.

Author Contributions: Designing the study, analyzing the data and writing the paper, E.-B.C.; Designing the study and supporting the writing of the manuscript related to resin collection history, C.-R.P.; Designing the study, analyzing the data and supporting the writing of the manuscript, J.-W.S.; Measuring the ring widths and cross-dating the tree-ring time series, Y.-J.K.; Sampling in the field and preparation of the samples for the ring-width measurement, J.-H.P. All authors have read and agreed to the published version of the manuscript.

Funding: This study was funded by the National Institute of Forest Science of Korea, grant number NIFOS FE 0100-2017-05.

Conflicts of Interest: The authors declare no conflict of interest.

References

1. Arnold, J.M.; Pérez, M.R. Framing the issues relating to non-timber forest products research. In *Current Issues in Non-timber Forest Product Research*; CIFOR/ODA: Bogor, Indonesia, 1996; pp. 59–81.
2. Park, C.; Lee, M.; Yoo, S.; Kim, S. *Conservation and Their Social Sharing of Pine Trees Damaged by Over Use of Pine Resin*; Research Report No.778; NIFOS: Seoul, Korea, 2018.
3. Gan, W. CHISENGYORAM. 1691, Volume 2. Available online: http://encykorea.aks.ac.kr (accessed on 1 November 2020).
4. Seo, M. GOSASINSEO. 1771, Volume 15. Available online: http://encykorea.aks.ac.kr (accessed on 1 November 2020).
5. Choi, H. NONJEONGHOIYO. 1831, Volume 10. Available online: http://encykorea.aks.ac.kr (accessed on 1 November 2020).
6. Stranges, A.N. Synthetic fuel production in prewar and world war II Japan: A case study in technological failure. *J. Ann. Sci.* **2006**, *50*, 3. [CrossRef]
7. Mi, K.Y. Wartime Munitions Securement and Korean Labor Mobilization pine resin turpentine, pine oil as oil substitute. *Hist. Assoc. Soong Sil* **2010**, *25*, 207–237.
8. Schweingruber, F. *Tree Rings—Basics and Applications of Dendrochronology (Dordrecht: Reidel)*; Springer: Berlin/Heidelberg, Germany, 1988.
9. Schmid, R.; Schweingruber, F.H. Tree Rings and Environment: Dendroecology. *Taxon* **1997**, *46*, 604. [CrossRef]
10. Bigler, C.; Bugmann, H. Growth-dependent tree mortality models based on tree rings. *Can. J. For. Res.* **2003**, *33*, 210–221. [CrossRef]
11. Stoffel, M.; Bollschweiler, M. Tree-ring analysis in natural hazards research—An overview. *Nat. Hazards Earth Syst. Sci.* **2008**, *8*, 187–202. [CrossRef]
12. Rybníček, M.; Kočár, P.; Muigg, B.; Peška, J.; Sedláček, R.; Tegel, W.; Kolář, T. World's oldest dendrochronologically dated archaeological wood construction. *J. Archaeol. Sci.* **2020**, *115*, 105082. [CrossRef]
13. Klein, A.; Nemestothy, S.; Kadnar, J.; Grabner, M. Dating furniture and coopered vessels without waney edge—Reconstructing historical wood-working in Austria with the help of dendrochronology. *Dendrochronologia* **2014**, *32*, 90–96. [CrossRef] [PubMed]
14. Stephen, E. Nash Archaeological Tree-Ring Dating at the Millennium. *J. Archaeol. Res.* **2002**, *10*, 243–275.
15. Čufar, K.; Beuting, M.; Demšar, B.; Merela, M. Dating of violins—The interpretation of dendrochronological reports. *J. Cult. Herit.* **2017**, *27*, S44–S54. [CrossRef]
16. Shindo, L.; Claude, S. Buildings and wood trade in Aix-en-Provence (South of France) during the Modern period. *Dendrochronologia* **2019**, *54*, 29–36. [CrossRef]
17. Vasileva, P.; Panayotov, M. Dating fire events in Pinus heldreichii forests by analysis of tree ring cores. *Dendrochronologia* **2016**, *38*, 98–102. [CrossRef]
18. Szymczak, S.; Bräuning, A.; Häusser, M.; Garel, E.; Huneau, F.; Santoni, S. A Dendroecological Fire History for Central Corsica/France. *Tree Ring Res.* **2020**, *76*, 40–53. [CrossRef]
19. Struble, W.T.; Roering, J.J.; Black, B.A.; Burns, W.J.; Calhoun, N.; Wetherell, L. Dendrochronological dating of landslides in western Oregon: Searching for signals of the Cascadia A.D. 1700 earthquake. *GSA Bull.* **2020**, *132*, 1775–1791. [CrossRef]
20. Ballesteros, J.; Stoffel, M.; Bodoque, J.; Bollschweiler, M.; Hitz, O.; Díez-Herrero, A. Changes in Wood Anatomy in Tree Rings of Pinus pinaster Ait. Following Wounding by Flash Floods. *Tree Ring Res.* **2010**, *66*, 93–103. [CrossRef]
21. Meko, M.D.; Therrell, M.D. A record of flooding on the White River, Arkansas derived from tree-ring anatomical variability and vessel width. *Phys. Geogr.* **2019**, *41*, 83–98. [CrossRef]
22. Lynch, A.M. What tree-ring reconstruction tells us about conifer defoliator outbreaks. In *Insect Outbreaks Revisited*; John Wiley & Sons: Hoboken, NJ, USA, 2012; pp. 126–154.
23. Kaennel, M.; Schweingruber, F.H. Multilingual Glossary of Dendrochronology. *WSL FNP Haupt* **1995**, *133*, 162–184.
24. Baillie, M.G.; Pilcher, J.R. A Simple Crossdating Program for Tree-Ring Research. *Tree Ring Bull.* **1973**, *33*, 7–14.

25. Eckstein, D.; Bauch, J. Beitrag zur Rationalisierung eines dendrochronologischen Verfahrens und zur Analyse seiner Aussagesicherheit. *Eur. J. For. Res.* **1969**, *88*, 230–250. [CrossRef]
26. Kwon, S.-M.; Kim, N.-H. Annual ring formation of major wood species growing in Chuncheon, Korea (I)-the period of cambium activity. *J. Korean Wood Sci. Technol.* **2005**, *33*, 1–8.
27. Park, S.-Y.; Eom, C.-D.; Seo, J.-W. Seasonal Change of Cambium Activity of Pine Trees at Different Growth Sites. *J. Korean Wood Sci. Technol.* **2015**, *43*, 411–420. [CrossRef]
28. Seo, J.-W.; Choi, E.-B.; Ju, J.-D.; Shin, C.-S. The Association of Intra-Annual Cambial Activities of Pinus koraiensis and Chamaecyparis pisifera planted in Mt. Worak with Climatic Factors. *J. Korean Wood Sci. Technol.* **2017**, *45*, 43–52. [CrossRef]
29. Funada, R.; Yamagishi, Y.; Begum, S.; Kudo, K.; Nabeshima, E.; Nugroho, W.D.; Hasnat, R.; Oribe, Y.; Nakaba, S. Xylogenesis in Trees: From Cambial Cell Division to Cell Death. In *Secondary Xylem Biology*; Elsevier BV: Amsterdam, The Netherlands, 2016; pp. 25–43.
30. Seo, J.-W.; Eckstein, D.; Schmitt, U. The pinning method: From pinning to data preparation. *Dendrochronologia* **2007**, *25*, 79–86. [CrossRef]
31. Bhatt, J.; Nair, M.; Ram, H.M. Enhancement of oleo-gum resin production in Commiphora wightii by improved tapping technique. *Curr. Sci.* **1989**, *58*, 349–357.
32. Shigo, A.L. Tree health. *J. Arboric.* **1982**, *8*, 311–316.

Publisher's Note: MDPI stays neutral with regard to jurisdictional claims in published maps and institutional affiliations.

Article

The Functional Traits of Breeding Bird Communities at Traditional Folk Villages in Korea

Chan Ryul Park *, Sohyeon Suk and Sumin Choi

Urban Forests Research Center, National Institute of Forest Science, 57, Hoegiro, Dongdaemun-gu, Seoul 02455, Korea; m17swannebula@gmail.com (S.S.); ciromi@korea.kr (S.C.)

* Correspondence: maeulsoop@korea.kr; Tel.: +82-2-961-2612

Received: 29 September 2020; Accepted: 9 November 2020; Published: 10 November 2020

Abstract: Interaction between nature and human has formulated unique biodiversity in temperate regions. People have conserved and maintained traditional folk villages (TFVs) dominated with houses made of natural materials, arable land and surrounding elements of landscape. Until now, little attention has been given to understand the traits of breeding birds in TFVs of Korea. The aim of this study was to reveal traits of breeding birds in TFVs and get conservative implications for biodiversity. We selected five TFVs: Hahoe maeul (HA), Wanggok maeul (WG), Nagan maeul (NA), Yangdong maeul (YD), and Hangae maeul (HG). We surveyed breeding birds with line transect methods, and analyzed functional traits (diet type and nest type) of birds in TFVs. Among 60 species recorded, *Passer montanus* (PM), *Streptopelia orientalis* (SO), *Hirundo rustica* (HR), *Pica pica* (PP), *Phoenicuros auroreus* (PA), *Paradoxornis webbiana* (PW), *Microscelis amaurotis* (MA), *Carduelis sinica* (CA) and *Oriolus chinensis* (OC) could be potential breeding birds that prefer diverse habitats of TFVs in Korea. Compared to the breeding birds of rural, urban and forest environments, the diversity of nesting types for birds was high in TFVs. The diverse nest types of breeding birds can be linked with habitat heterogeneity influenced by sustainable interaction between nature and human in TFVs in Korea.

Keywords: backyard forest; functional traits; livelihood; nesting guild; pungsu

1. Introduction

Since the Rio Declaration, Agenda 21 in Brazil of 1992, international attentions have been given to indigenous people at tropic regions; a number of studies have been conducted to comprehend and suggest the importance of livelihood and biodiversity of local communities. However, there are a couple of well-known or sightseeing sites of traditional folk villages (TFVs) at temperate regions which have intentionally been conserved and maintained by government supports [1–3]. We need to understand the relationship between biodiversity and TFVs at typical monsoon lifestyles of agricultural and forestry cultivating systems in China, Japan and Korea. Biocultural diversity is known to be connected with the interrelationship between dwellers and nature [4], and also with functional roles of habitats, especially from cultural and socio-economic perspectives [5–8]. Tropic regions have high biodiversity in residential areas, while temperate regions have a high biodiversity by products of a long-harmonized relationship between humans and nature. In 1945, German geographer Lautensach [9] commented on the pattern of Korean settlement in comparison with ASEAN (Association of South-East Asian Nations) people. He pointed out the diverse spectrum from slash-and-burn farming to residential agriculture in Korea, contrary to the ASEAN peoples showing the migrating and turning patterns of slash-and-burn farming. Based on his records, Korea showed permanent settled or migrating patterns of slash-and-burn farming at some areas.

However, there are a few villages conserving old lifestyles and residential patterns in Korea. With changing climate and socio-economic conditions, it is very difficult to maintain TFVs on their own finance, so financial supports by central and local government are essential to conserve folk villages and typical residential patterns like thatched houses and tile-roofed houses. Park (2008) [10] suggested that the biodiversity of a rural landscape can be represented by three kinds of interactions like edge effects, landscape complementation and mutual synchronization. However, there has been little survey on the traits of breeding birds in TFVs. This study was conducted to reveal the traits of breeding birds and find a conservation measure in TFVs in Korea.

2. Materials and Methods

2.1. Study Site

We selected five TFVs based on the settlement of villagers and maintenance of traditional residential circumstances like thatched houses and tile-roofed houses. Maeuls refer to the village in pure Korean language. Among the five villages, three villages (Hahoe, Yangdong and Hangae maeuls) belong to Gyeonsangbukdo province. Wanggok maeul is located at the most northern parts, while Nagan maeul is at the most southern parts of Korean peninsula. The size of Hahoe maeul amounts as 904,821 square meters; the others are below 530,000 square meters. Based on the size of survey areas, we chose survey distance (Figure 1, Table 1).

2.2. Bird Survey Method

By using the digital contour maps (1:25,000 or 1:5000) and forest cover maps (1:5000), survey routes were chosen to include all habitats of birds. The name and area of each site were referred to the regional legends of local government. Birds were surveyed three times in each site with the line transect method in the morning from May 1 to July 5 in 2016. To avoid bias from repeated observations of the same individuals, we surveyed birds while walking at the speed of 2 km per hour between 0530 and 0800 on a clear day. Census trails were set up at the length enough to determine the number of species present in each site. All birds seen or heard within 25 m either side of the census trail were identified by song, call, flying type and field mark by eye or with binoculars (8 × 30). All birds seen were recorded and identified by binoculars, song and call, and the number of individuals were counted; the density was calculated as an individual density(ea/km/hr) of each species [11,12].

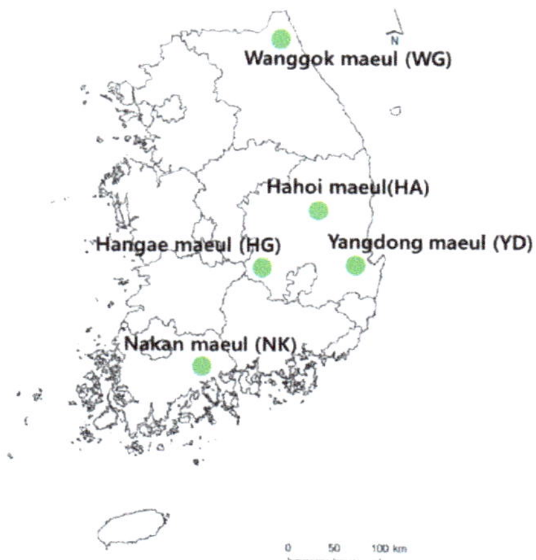

Figure 1. Five traditional folk villages (TFVs) were selected based on the villagers' settlements and maintenance of traditional residential circumstances.

2.3. Habitat Survey Method

In the viewpoint of bird's habitat, the elements of landscapes at traditional folk village include diverse patches like houses (thatched or tile-roofed), cultivation areas (croplands, paddies and orchards), wetlands (rivers and ponds) and forests. By using the digital contour maps (1:25,000 or 1:5000) and forest cover maps (1:5000), we classified land cover into 13 categories, including paddies, deciduous forests, grasslands, coniferous forests, croplands, rivers, bare lands, residential areas, mixed forests, riparian shrubs, roads, bamboo forests and ponds (Table 2). After classification of land cover, we calculated the percentage value of each cover with the application of ArcMap 9.3 (ESRI, 1999). Habitat diversity and diversity of functional traits were calculated using the Shannon-Wiener diversity index:

$$H' = -\text{pi}\sum_{i=1}^{n} \ln(pi) \quad (1)$$

Table 1. Locations, areas and survey distances at five maeuls.

Maeuls * (Villages)	Abbreviations	Locations	Area (m^2)	Survey Distance (km)
Hahoe	HA	Andong City, Gyeongsangbukdo	904,821	2
Wanggok	WG	Goseong County, Gangwondo	497,560	1
Nagan	NA	Suncheon City, Jeollanamdo	317,548	1
Yangdong	YD	Gyeongju City, Gyeongsangbukdo	392,976	1
Hangae	HG	Seongju City, Gyeongsangbukdo	527,333	1

* Maeul refers to village in pure Korean language.

Table 2. Percentage of each land cover and habitat diversity index at five maeuls.

Land Cover	HA *	WG *	NA *	YD *	HG *
Paddies	32.9	5.1	27.1	1.1	3.2
Deciduous forests	0.4	33.8	1.8	22.8	24.2
Grasslands	10.2	4.8	27.9	34.6	9.5
Coniferous forests	1.9	19.8	0.0	15.2	27.3
Croplands	8.7	9.9	12.8	6.6	8.7
Rivers	17.7	0.6	1.7	0.9	0.0
Bare lands	4.9	7.8	10.7	3.4	3.2
Residential areas	5.1	2.0	9.2	8.2	3.1
Mixed forests	0.0	8.8	0.0	0.0	15.5
Riparian shrubs	13.8	0.0	0.0	0.0	0.0
Roads	4.0	3.5	6.0	7.3	2.9
Bamboo forests	0.0	3.8	2.2	0.0	2.4
Ponds	0.5	0.0	0.5	0.0	0.0
Total (m^2)	904,820.7	497,560.0	317,547.6	392,976.3	527,332.7
Habitat diversity	1.9390	1.9800	1.8522	1.7713	1.9444

* HA: Hahoe, WG: Wanggok, NA: Nagan, YD: Yangdong, HG: Hangae.

2.4. Guild Analysis and Functional Guild

We applied functional traits concept to comprehend the nesting resources of breeding bird community, classified into nesting type (hole, canopy, bush (including ground), house and water) and diet type (granivores, insectivores, omnivores, piscivores, predators, scavengers, shorebirds and riverine insectivores) [13,14].

2.5. Ordination Analysis and Simple Regression

We conducted non-multidimensional scaling methods to compare the densities of bird communities among five maeuls with the Past 3.13 program [15]. We compared the number of birds and density with habitat diversity with a simple general linear model. We conducted the arcsine transformation of each ratio of land cover for normal distribution of variables.

3. Results

3.1. Characteristics of Breeding Bird Community at Traditional Folk Villages

We recorded total 60 species of birds at five maeuls, including 20 canopy nesters, 13 hole nesters, 8 bush nesters, 7 water nesters and 5 house nesters. From a viewpoint of diet type, the species were mainly composed of 35 insectivores, 8 predators and 6 piscivores (Appendix A). Among the five regions, Wanggok (WG) maeul showed the highest number of birds, 38 species, and Hangae maeul had the lowest value, 29 species of birds. Hole, bush and canopy nesters were highly observed at Wanggok maeul (WG), while nesting birds near water areas were high at Hahoe maeul (Figure 2a). From the viewpoint of average densities, the values of house nesters were high at Yangdong, Nagan, Wanggok and Hahoe maeul, but low at Hangae maeul (Figure 2b). Diversity of diet types showed the highest value of 1.4967 at Hahoe maeul, and the lowest of 1.2968 at Nagan maeul. Densities of riverine insectivores like wagtails and plovers were high at Hahoe maeul (Figure 2c)

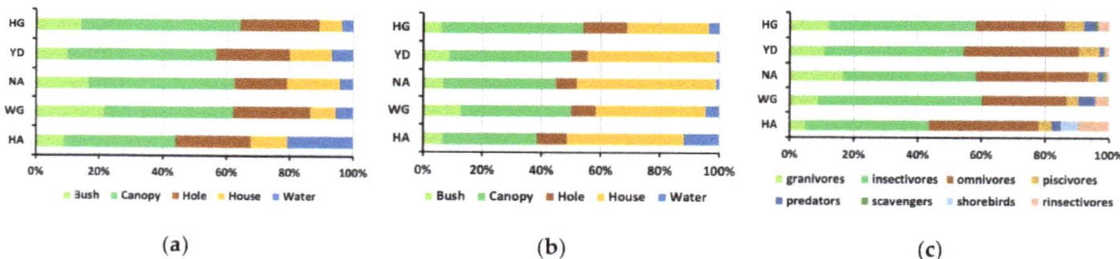

Figure 2. Percentage of nesting type based on the number of species (**a**), average densities (**b**) and diet types based on the average densities (**c**) at five maeuls.

3.2. Comparison among Sites and Observed Species

Ordination analysis indicated that Hangae (HG) and Wanggok (WG) maeul showed similarity in the characteristics of bird community; however, the other three sites showed relative unique patterns of bird community (Figure 3). Two sites were composed of high value of forests, 62.4% in WG and 67.0% in HG, so forest birds could dominate the bird community. Hahoe (HA) maeul is located near the river, so riverine birds occurred more highly than at the other sites. Yangdong (YD) maeul showed high percentage of grasslands, and Nakan (NA) maeul a high percentage of bare lands. Thus, among the five TFVs, WG and HG maeuls showed similarity in bird species composition, but the others reflected the characteristics of habitat type for their birds.

Among the observed 60 species, nine species (Passer montanus, Streptopelia orientalis, Hirundo rustica, Pica pica, Phoenicuros auroreus, Paradoxornis webbiana, Microscelis amaurotis, Carduelis sinica and Oriolus chinensis) showed increasing dissimilarity in relation to the other birds which aggregated similarity at the red-dotted circle based on the average density of the five maeuls (Figure 4). Within the red-dotted circle, the birds mostly belong to the forest-dwelling birds. Thus, we could infer that the nine species are related with habitat types of TFVs. Among the nine species, Hirundo rustica and Oriolus chinensis were summer visitors which migrate to Southeast Asia to spend winter, and seven birds were residents.

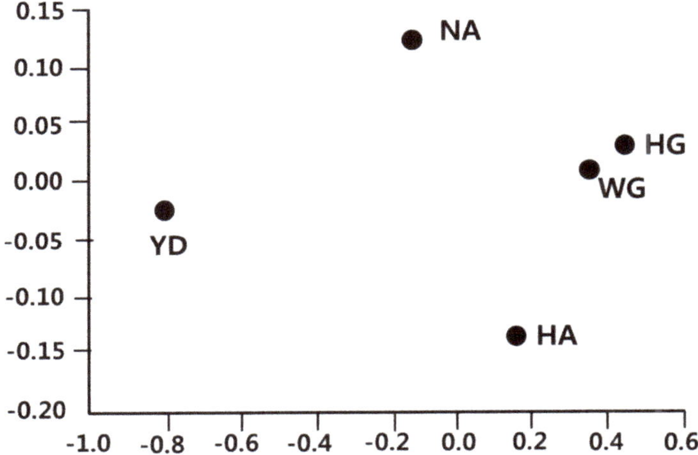

Figure 3. Non-multidimensional scaling ordination with the average density of observed birds at five maeuls (YD-Yandong, NA-Nagan, HA-Hahoe, WG-Wangok, HG-Hangae; Stess-0.0, R2-0.9967) by Past, Program V1.35b.

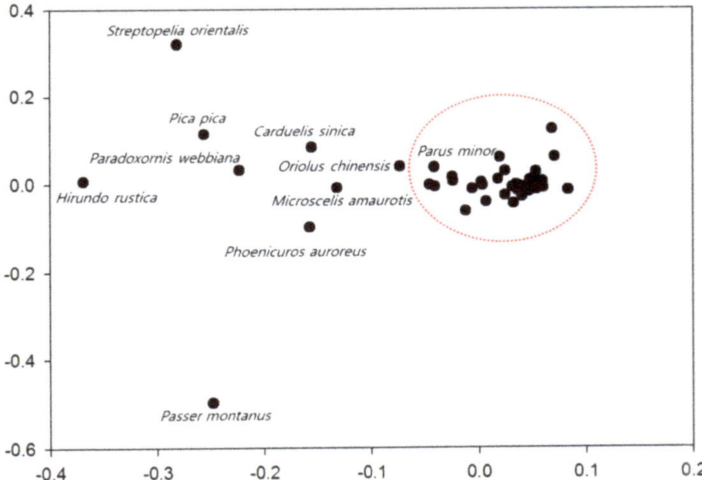

Figure 4. Non-multidimensional scaling ordination with the average density of sixty observed birds at five maeuls by Past, Program V1.35b (Red dotted circles include forest-dwelling birds, Stress-0.13, R^2-0.7962).

3.3. *Relationship Habitat Diversity and Number of Birds*

At the five traditional folk villages, habitat diversity highly influenced nesting type diversity (R^2 = 0.74, Figure 5a), and has a weak relationship with diet type diversity (R^2 = 0.35, Figure 5b) and overall number of birds' species (R^2 = 0.31, Figure 5c). These results indicate that habitat heterogeneity of traditional rural landscape can provide diverse nesting resources for birds and indirectly affect the functional groups such as insectivores, granivores and predators.

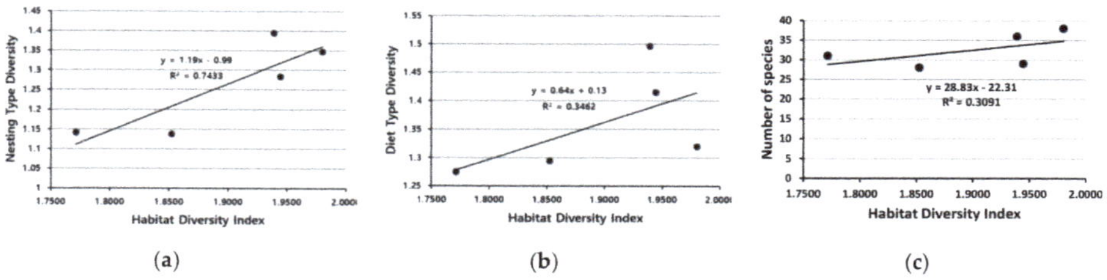

(**a**) (**b**) (**c**)

Figure 5. Relationships between habitat diversity index and nesting type diversity (**a**), diet type diversity (**b**) and number of species (**c**) at study sites.

4. Discussion

4.1. The Traits of Breeding Bird Communities at Traditional Folk Villages in Korea

TFVs possess diverse habitats like houses (thatched or tile-roofed), cultivation areas (croplands, paddies and orchards), wetlands (rivers and ponds) and forests for breeding birds. Nine species showed different occurrence patterns with forest-dwelling birds at five maeuls in Korea. This could suggest that nine species (*Passer montanus* (PM), *Streptopelia orientalis* (SO), *Hirundo rustica* (HR), *Pica pica* (PP), *Phoenicuros auroreus* (PA), *Paradoxornia webbiana* (PW), *Microscelis amaurotis* (MA), *Carduelis sinica* (CA) and *Oriolus chinensis* (OC)) can be potential species that prefer diverse habitats of TFVs in Korea. Five out of the nine species were canopy nesters (SO, PP, MA, CS, OC), three were house nesters (HR, PA, PM) and one was a bush nester (PW) according to the nest types. These species reflect the diverse use of nesting resources such as trees, shrubs and traditional thatched house in TFVs. Diversity index of nesting type was higher in the order of TFVs, rural [16], urban [12] and forests [17] (Table 3). Diversity of nesting type could be related with functional traits of breeding bird communities in TFVs.

Table 3. Comparison of diversity of nesting types among different systems in Korea.

Sources	Hole	Canopy	Bush	House	Water	Diversity Index	# of Sites
TFVs	13	20	8	5	7	1.4880	5
Rural [16]	7	10	7	3	-	1.3165	43
Urban [12]	10	13	8	-	-	1.0790	2
Forests [17]	9	5	5	-	-	1.0584	6

4.2. Residential Houses of Nesting Resource for Birds

Among the three house nesters, PM make nests at the multi-porous space of a thatched house with the lining resources of straw of herbs Gramineae and Cyperaceae [18], HR make nests with the muds and straws beneath parts of roof at thatched and tile-roofed house [19], and PA utilize the needles of pines and herbs for lining resources at nests in thatched and tile-roofed house [20]. This represents that the people and birds do coexist and assist breeding success against harsh climate condition and predators. HR was known to prefer the human-dominated house to lessen the predation risk from cuckoos [21]. Thus, residential houses for local people can provide breeding nests for birds in TFVs in Korea. Long-time interaction networks among humans, birds and plants [10] can affect the specific composition of bird community in TFVs in Korea.

4.3. Sustainability of Breeding Birds at Traditional Folk Villages in Korea

In Korea, urbanization and industrialization have impacted rural society since the 1970s [22], while TFVs conserve and protect traditional rural landscapes and high biodiversity to this day. Recently TFVs have been cited as pilot villages for low-carbon cities and sustainable urban environments due to COVID-19 and climate change. From this viewpoint, the value and importance of biodiversity should be evaluated as a new perspective for nature-based solutions like natural healing resources in TFVs. Meanwhile, urban shrinking and climate change could threat the sustainability of nesting & diet types in TFVs in Korea. Recently, biological conservation implication as well as socio-economic policy for stakeholders in rural landscapes could protect and conserve the biodiversity at a specific city and rural community [23–25].

Meanwhile, two summer visitors (HR, OC) do breed in TFVs in Korea, but they migrate to the Southeast Asia to spend winter. To conserve the population of these birds, it is important to ensure wintering grounds and staging grounds at the migration routes in East Asia. Also, it is necessary to assess the habitat of two birds at breeding grounds of Korea and wintering grounds of Southeast Asia. The application of a city biodiversity index [26] would be recommended to evaluate habitats and the biodiversity in TFVs in the Asia region.

Compared to rural, urban and forest habitats, TFVs showed a high diversity of nesting types for breeding birds. The functional traits of breeding birds in TFVs can be related with habitat heterogeneity or simultaneous use by breeding birds at forests, paddies and croplands.

Two cases of the western Amazon [27] and Australian region [28] showed that the importance of biodiversity and livelihoods of indigenous people should be considered before environmental impact assessment of government development policies. From this viewpoint, the relationship between functional traits of breeding birds and the lifestyle of TFVs should be more addressed and interpreted to comprehend the value of TFVs in Korea.

Local & central government have endeavored to conserve TFVs with legislative and economic support to the villagers; however, in-depth excavation of value of TFVs in terms of biodiversity and economic valuation should be executed in the near future. TFVs can provide a harmonized solution of valuable nature and curable nurture in a post-COVID-19 society.

5. Conclusions

Villagers in TFVs depend on their livelihoods such as rice production and greens cultivation in agricultural fields, gathering wild edible green and mushroom production in forests. They live in their thatched or roof-tiled houses where swallows and sparrows can safely make breeding nests against predators. Their eco-friendly livelihoods, transcended from old times, allow them to interact with birds, and their harmonized attitudes can enhance biodiversity in TFVs. However, socio-economic change and individualized society can adversely affect the bird diversity as well as local villagers. Research on the interaction between local people and biodiversity should be conducted to sustain TFVs against climate change and COVID-19.

Author Contributions: S.S. and C.R.P. collected the field data and wrote the paper; S.S. and C.R.P. collected the field data and analyzed the statistical data on geographic information data; S.C. managed the research project and contributed to the discussion on data; C.R.P. designed the study site and surveyed plans and statistical analyses of all data. All authors have read and agreed to the published version of the manuscript.

Funding: This study was funded by National Institute of Forest Science of Korea, grant number NIFOS FE0100-2017-05.

Conflicts of Interest: The authors declare no conflict of interest.

Appendix A

Table A1. Densities (ea/km/hr) of bird species included in this study and their functional traits. RInsectivore = riverine insectivore.

Scientific Name	Study Sites					Nesting Type	Diet Type
	HA	WG	NA	YD	HG		
Accipiter gentilis	0.0	0.0	0.3	0.0	0.0	Canopy	Predator
Accipiter soloensis	0.7	0.0	0.0	0.3	0.0	Canopy	Predator
Actitis hypoleucos	0.3	0.0	0.0	0.0	0.0	Water	Shorebird
Aix galericulata	2.3	0.0	0.0	1.0	0.0	Hole	RInsectivore
Alcedo atthis	0.0	0.0	0.3	0.0	0.0	*	Piscivore
Anas platyrhynchos	0.3	0.0	0.0	0.0	0.0	Water	RInsectivore
Anas poeilorhyncha	0.7	0.0	0.0	0.7	1.3	Water	RInsectivore
Ardea cinerea	1.3	0.0	1.0	3.0	1.0	Canopy	Piscivore
Butorides striatus	0.0	0.0	0.3	2.3	0.3	Canopy	Piscivore
Caprimulgus indicus	0.3	0.0	0.0	0.7	0.0	Bush	Insectivore
Carduelis sinica	2.0	0.3	0.7	8.0	0.0	Canopy	Insectivore
Cettia diphone	0.0	2.7	0.0	0.0	0.0	Bush	Insectivore
Charadrius dubius	2.0	0.0	0.0	0.0	0.0	Water	Shorebird
Charadrius placidus	1.0	0.0	0.0	0.0	0.0	Water	Shorebird
Corvus macrorhynchos	0.0	0.0	0.7	0.0	0.3	Canopy	Scavenger
Cuculus canorus	1.7	1.3	1.0	2.0	0.7	*	Insectivore
Cuculus micropterus	0.7	0.0	0.3	0.0	0.0	*	Insectivore
Cuculus poliocephalus	0.0	0.0	1.0	0.0	0.0	*	Insectivore
Cyanopica cyana	0.0	5.0	0.0	0.0	0.0	Canopy	Insectivore
Cyanoptila cyanomelana	0.0	0.0	0.0	0.0	0.7	Canopy	Insectivore
Dendrcopos major	0.0	0.7	0.0	0.7	1.7	Hole	Insectivore
Dendrocopos kizuki	0.3	0.0	0.0	0.7	0.7	Hole	Insectivore
Dendrocopos leucotos	0.0	0.0	0.0	0.0	0.3	Hole	Insectivore
Dendrocopos spp.	0.3	0.0	0.0	0.0	0.0	*	*
Egretta alba	1.3	0.0	0.0	2.7	0.7	Canopy	Piscivore
Egretta intermedia	0.0	0.0	0.0	0.0	0.3	Canopy	Piscivore
Egretta spp.	0.0	0.7	0.3	0.0	0.0	*	*
Eurystomus orientalis	1.7	0.3	1.0	2.7	1.0	Canopy	Insectivore
Falco subbuteo	0.0	0.3	0.0	0.3	0.7	Canopy	Predator
Falco tinnunculus	0.0	0.0	0.0	0.7	0.0	House	Predator
Garrulus glandarius	0.0	0.0	0.0	0.3	0.0	Canopy	Omnivore
Halcyon coromanda	0.0	1.7	0.3	0.3	0.0	Hole	Piscivore
Hirundo daurica	0.3	0.0	0.0	0.0	0.0	House	Insectivore
Hirundo rustica	4.7	1.7	8.0	12.7	0.0	House	Insectivore
Lanius bucephalus	0.0	0.7	0.7	0.0	0.7	Bush	Predator
Lanius cristatus	0.0	0.7	0.0	0.0	0.0	Bush	Predator
Lanius tigrinus	0.0	0.7	0.0	0.0	0.0	Bush	Predator
Microscelis amaurotis	0.0	2.0	3.7	5.3	1.3	Canopy	Insectivore
Motacilla alba	2.0	2.0	0.7	0.3	0.0	Water	RInsectivore
Motacilla grandis	0.7	0.0	0.0	0.0	0.0	Water	RInsectivore
Motacilla spp.	0.0	0.0	0.3	0.0	0.0	*	*
Oriolus chinensis	1.7	1.3	1.3	4.0	2.0	Canopy	Insectivore
Otus scops	1.0	0.0	0.0	0.3	0.0	Hole	Predator
Paradoxornis webbiana	2.3	0.3	3.0	9.3	1.0	Bush	Insectivore
Parus ater	0.0	0.0	0.0	0.0	0.3	Hole	Insectivore
Parus minor	0.7	0.7	1.3	2.7	2.0	Hole	Insectivore
Parus palustris	0.3	0.0	0.0	0.0	0.0	Hole	Insectivore
Passer montanus	14.3	9.7	19.0	36.7	6.7	House	Omnivore
Phasianus colchicus	1.3	0.7	0.7	1.0	0.3	Bush	Insectivore
Phoenicuros auroreus	4.3	5.0	2.7	3.3	3.7	House	Insectivore
Phylloscopus occipitalis	0.0	0.0	0.0	0.0	0.3	Bush	Insectivore
Pica pica	7.0	2.3	4.0	7.3	4.0	Canopy	Omnivore
Picus canus	0.0	0.3	0.0	0.0	0.3	Hole	Insectivore

Table A1. *Cont.*

Scientific Name	Study Sites					Nesting Type	Diet Type
	HA	WG	NA	YD	HG		
Sitta europaea	0.0	0.0	0.0	0.0	0.3	Hole	Insectivore
Streptopelia orientalis	3.0	4.0	11.0	13.3	4.7	Canopy	Granivore
Sturnus cineraceus	1.3	0.0	2.7	1.0	0.0	Hole	Insectivore
Sturnus philippensis	0.0	0.3	0.0	0.0	0.0	Hole	Insectivore
Turdus hortulorum	0.0	0.7	0.0	0.3	0.7	Canopy	Insectivore
Turdus pallidus	0.0	0.0	0.0	0.0	0.3	Canopy	Insectivore
Zoothera dauma	0.3	0.0	0.0	0.0	0.0	Canopy	Insectivore

Study sites: Hahoe maeul (HA), Wanggok maeul (WG), Nagan maeul (NA), Yangdong maeul (YD), and Hangae maeul (HG). * Species were omitted for guild characterization due to the peculiarity of breeding habit or non-breeders.

References

1. Kohsaka, R.; Tomiyoshi, M.; Saito, O.; Hashimoto, S.; Mohammend, L. Interactions of knowledge systems in shiitake mushroom production: a case study on the Noto Peninsula, Japan. *J. For. Res.* **2015**, *20*, 453–463. [CrossRef]
2. Tashiro, A.; Uchiyama, Y.; Kohsaka, R. Impact of Geographical Indication schemes on traditional knowledge in changing agricultural landscapes: An empirical analysis from Japan. *J. Rural Stud.* **2019**, *68*, 46–53. [CrossRef]
3. Kohsaka, R.; Sun, M.; Uchiyama, Y. Beekeeping and honey production in Japan and South Korea: past and present. *J. Ethn. Foods* **2017**, *4*, 72–79. [CrossRef]
4. Parrotta, J.A.; Trosper, R.L. Traditional Forest-related Knowledge: Sustaining Communities. In *Ecosystem and Biocultural Diversity*; Springer: London, UK, 2012; p. 621.
5. Balmford, A.; Moore, J.L.; Brooks, T.; Burgess, N.; Hansen, L.A.; Williams, P.; Rahbek, C. Conservation conflicts across Africa. *Science* **2001**, *291*, 2616–2619. [CrossRef]
6. Bodin, Ö.; Tengö, M.; Norman, A.; Lundberg, J.; Elmqvist, T. The value of small size: Loss of forest patches and ecological thresholds in southern Madagascar. *Ecol. Appl.* **2006**, *16*, 440–451. [CrossRef]
7. Herrera, J.M.; García, D. The role of remnant trees in seed dispersal through the matrix: Being alone is not always so sad. *Biol. Conserv.* **2009**, *142*, 149–158. [CrossRef]
8. Lütz, M.; Bastian, O. Implementation of landscape planning and nature conservation in the agricultural landscape—A case study from saxony. *Agric. Ecosyst. Environ.* **2002**, *92*, 159–170. [CrossRef]
9. Lautensach, H. *Korea: Eine Landeskunde auf Grund Eigener Reisen und der Literatur*; KF Koehler: Leipzig, Germany, 1945.
10. Park, C.-R. Interaction networks of organisms at Korean rural landscape. In *Traditional Ecology of Korea 2nd Volume: Interaction Networks of Organisms at Korean Rural Landscape*; Lee, D., Ed.; Sciencebooks: Seoul, Korea, 2008; p. 463.
11. Bibby, C.J.; Burgess, N.D.; Hill, D.A.; Mustoe, S. *Bird Census Techniques*; Elsevier: Amsterdam, The Netherlands, 2000.
12. Park, C.-R.; Lee, W.-S. Relationship between species composition and area in breeding birds of urban woods in Seoul, Korea. *Landsc. Urban Plan.* **2000**, *51*, 29–36. [CrossRef]
13. Lee, W.-S.; Park, C. Analysis of Changes on the Forest Environment and the Bird Community in Terms of "Guild". *Korean J. Ecol.* **1995**, *18*, 397–408.
14. Martin, E.A.; Seo, B.; Park, C.R.; Reineking, B.; Steffan-Dewenter, I. Scale-dependent effects of landscape composition and configuration on natural enemy diversity, crop herbivory, and yields. *Ecol. Appl.* **2016**, *26*, 448–462. [CrossRef]
15. Hammer, Ø.; Harper, D.A.T.; Ryan, P.D. PAST: Paleontological statistics software package for education and data analysis. *Palaeontol. Electron.* **2001**, *4*, 9.
16. Park, C.; Lee, W. Effects of fragmentation on the bird community in agricultural landscapes. *Korean J. Environ. Ecol.* **2002**, *16*, 22–32.
17. Park, C. Tree species preference and inter-specific difference of foraging maneuver, trees and location among four canopy-dwelling birds at high-elevation temperate deciduous forest in Mt. Jumbongsan. *Integr. Biosci.* **2005**, *9*, 41–46. [CrossRef]

18. Field, R.H.; Anderson, G.Q.A.; Gruar, D.J. Land-use correlates of breeding performance and diet in Tree Sparrows Passer montanus. *Bird Study* **2008**, *55*, 280–289. [CrossRef]
19. Osawa, T. Importance of farmland in urbanized areas as a landscape component for barn swallows (Hirundo rustica) nesting on concrete buildings. *Environ. Manag.* **2015**, *55*, 1160–1167. [CrossRef]
20. Lee, W.-S.; Goo, T.; Park, J. *LG Green Foundation*; Minsokwon: Seoul, Korea, 2005; p. 383.
21. Liang, W.; Yang, C.; Wang, L.; Møller, A.P. Avoiding parasitism by breeding indoors: cuckoo parasitism of hirundines and rejection of eggs. *Behav. Ecol. Sociobiol.* **2013**, *67*, 913–918. [CrossRef]
22. Yu, D.J.; Anderies, J.M.; Lee, D.; Perez, I. Transformation of resource management institutions under globalization: The case of songgye community forests in South Korea. *Ecol. Soc.* **2014**, *19*. [CrossRef]
23. Uchiyama, Y.; Kohsaka, R. Application of the City Biodiversity Index to populated cities in Japan: Influence of the social and ecological characteristics on indicator-based management. *Ecol. Indic.* **2019**, *106*, 105420. [CrossRef]
24. Döringer, S.; Uchiyama, Y.; Penker, M.; Kohsaka, R. A meta-analysis of shrinking cities in Europe and Japan. Towards an integrative research agenda. *Eur. Plan. Stud.* **2020**, *28*, 1693–1712. [CrossRef]
25. Kohsaka, R. Developing biodiversity indicators for cities: Applying the DPSIR model to Nagoya and integrating social and ecological aspects. *Ecol. Res.* **2010**, *25*, 925–936. [CrossRef]
26. Uchiyama, Y.; Hayashi, K.; Kohsaka, R. Typology of Cities Based on City Biodiversity Index: Exploring Biodiversity Potentials and Possible Collaborations among Japanese Cities. *Sustainability* **2015**, *7*, 14371–14384. [CrossRef]
27. Finer, M.; Jenkins, C.N.; Pimm, S.L.; Keane, B.; Ross, C. Oil and gas projects in the Western Amazon: Threats to wilderness, biodiversity, and indigenous peoples. *PLoS ONE* **2008**, *3*. [CrossRef] [PubMed]
28. Bliege Bird, R.; Bird, D.W.; Codding, B.F.; Parker, C.H.; Jones, J.H. The "fire stick farming" hypothesis: Australian Aboriginal foraging strategies, biodiversity, and anthropogenic fire mosaics. *Proc. Natl. Acad. Sci. USA* **2008**, *105*, 14796–14801. [CrossRef] [PubMed]

Review

Subjective Well-Being as a Potential Policy Indicator in the Context of Urbanization and Forest Restoration

Takuya Takahashi [1,2,*], Yukiko Uchida [3], Hiroyuki Ishibashi [2,4] and Noboru Okuda [2,5]

1 School of Environmental Science, University of Shiga Prefecture, Hikone 522-8533, Japan
2 Research Institute for Humanity and Nature, Kyoto 603-8047, Japan; ishibashi_npo@aoni.waseda.jp (H.I.); nokuda@people.kobe-u.ac.jp (N.O.)
3 Kokoro Research Center, Kyoto University, Kyoto 606-8501, Japan; uchida.yukiko.6m@kyoto-u.ac.jp
4 Faculty of Human Sciences, Waseda University, Tokorozawa 359-1192, Japan
5 Research Center for Inland Seas, Kobe University, Kobe 657-8501, Japan
* Correspondence: tak@ses.usp.ac.jp or taka.takuya@gmail.com

Abstract: The enhancement of human well-being is one of the ultimate goals of resource management; however, it is not explicitly considered by forest policy indicators. Our previous studies examined how Japanese citizens in the Yasu River watershed of the Shiga Prefecture perceived subjective well-being related to forests (forest SWB). We found a negative correlation between forest SWB and forest ownership, suggesting dissatisfaction with the low profitability of forest ownership. Based on this result, in this paper, we argue that forest SWB can be an important indicator for policymaking in the context of urbanization and forest restoration and can complement existing forest indicators focusing mainly on physical and objective properties. First, we propose that a direct measurement of well-being (e.g., forest SWB) is preferable over an indirect measurement (e.g., GDP), for policymaking processes related to forests. Second, forest SWB can reflect the quality of our interactions with forests, which is important in urbanized societies which tend to have reduced experiences with nature. Third, forest SWB could identify inequalities between the users of forest ecosystem services and forest managers. Overall, forest SWB can be a holistic indicator to capture a variety of perspectives held by citizens.

Keywords: subjective well-being; happiness; policy indicator; forest policy; Japan

Citation: Takahashi, T.; Uchida, Y.; Ishibashi, H.; Okuda, N. Subjective Well-Being as a Potential Policy Indicator in the Context of Urbanization and Forest Restoration. *Sustainability* **2021**, *13*, 3211. https://doi.org/10.3390/su13063211

Academic Editors: Marc A. Rosen and Ryo Kohsaka

Received: 14 December 2020
Accepted: 10 March 2021
Published: 15 March 2021

Publisher's Note: MDPI stays neutral with regard to jurisdictional claims in published maps and institutional affiliations.

1. Introduction

1.1. Background

Urbanization is an ongoing and persistent global trend [1]. In 1950, 30% of the global population resided in urban areas. By 2018, this percentage increased to 55% and is expected to increase to 68% by 2050. This evidence has several policy implications because it poses environmental, economic, and social challenges for sustainable future development. The trend towards urbanization is also present in Japan. In 2018, 91.6% of the Japanese population resided in urban areas [1], ranking 17th in the world.

Forest restoration is also a global trend. As stated by Mather in his "forest transition hypotheses", forest areas generally decrease in beginning stages of economic development and, in turn, they increase in later developed stages [2–4]. Following intensive exploitation during World War II, Japanese forests have been restored and have now reached "forest saturation" status—a state where a nation has a sufficient quantity of forests [5].

In general, policy indicators determine the extent to which the policy goals are met and often reflect political or institutional factors that influence the entire process [6]. Given the connection between urbanization and forest restoration, forest policymakers need to consider new types of policy indicators that clarify the links between the subjective feelings of citizens and the condition of the forests. For example, following urbanization, the direct and material dependence on natural ecosystems has decreased, but the importance of spiritual (nonmaterial) values of natural ecosystems has increased [7].

Historically, policy indicators have tended to focus on the physical and objective conditions of the forest. For example, the Japanese national forest plan approved by the Cabinet office, Government in Japan (2018; Table 1) provided numerical goals for the types of forests and their areas, harvesting volumes, tree-planting areas, construction of forest roads, areas of protected forests, and soil conservation projects [8]. These physical forest characteristics have a significant impact on determining the amount of timber as well as the opportunities for recreation. For example, plantation forests and seminatural forests provide different levels of timber production and recreational opportunities. However, with such objective measures, consequences for the subjective well-being (SWB) of individuals who interact with forests remain unclear. Understanding these more subjective measures may help to address forest management problems commonly faced by policymakers in developed countries. For example, forest owners who lose interest in forest management do not properly manage forests [9,10].

Table 1. Structure of National Forest Plan (approved by the Cabinet on 16 October 2018) [8].

Chapters	Subchapters
I. Goals and other basic issues	1 Principles of forest management and protection 2 Goals of forest management and protection
II. Forest management	1 Harvesting/planting and thinning/tending 2 Publicly beneficial forests 3 Forest road construction and transportation of forest products 4 Rationalization of forest management
III. Forest protection	1 Forest land protection 2 Facilities 3 Protection from pests and fires
IV. Improvement of health-enhancing function of forests	1 Principles of setting up forests for health-enhancing function 2 Principles of management of forests for health-enhancing function 3 Other necessary issues

At the global level, the Food and Agricultural Organization of the United Nations (FAO) conducts a forest resource assessment every 5 years (Table 2) [11]. Nearly all items used in this series of assessments emphasizes forests' physical features, such as total forest areas and protected forest areas.

Table 2. Forest resources assessment (FRA) 2015 analysis data table [11].

TOPIC/Variable
Extent Area
Forest area Other wooded land Other land of which with tree cover
Forest Characteristics
Primary forest Other naturally regenerated forest Planted forest Area of mangrove forest

Table 2. *Cont.*

TOPIC/Variable
Growing Stock, Biomass and Carbon
Forest growing stock
Above-ground biomass
Below-ground biomass
Dead wood
Carbon in above-ground biomass
Carbon in below-ground biomass
Carbon in dead wood
Carbon in litter
Soil carbon
Production and Multiple use
Production forest
Multiple use forest
Biodiversity and Protected Areas
Conservation of biodiversity
Forest area within protected areas
Ownership of Forests
Public ownership
Private ownership
Unknown ownership
Management Rights of Public Forests
Public administration
Individuals
Private companies
Communities
Other
Employment in Forestry
Employment in forestry

1.2. Subjective Well-Being as a Policy Indicator

Subjective well-being (SWB) is a theoretical construct developed in psychology [12–14] and economics [15–17] and has been the subject of many empirical studies. SWB is a multidimensional construct capturing basic human psychological needs, such as security, basic materials for a satisfactory life, health, successful social relationships, and freedom of choice and action. These human needs are also of interest to the conceptual model of ecosystem services [18], and we argue it is therefore important to monitor well-being when considering policies for forest ecosystems.

Forest SWB refers to a subjective well-being measure that assesses the respondent's relationship with forests. Previous studies regarding forest-specific SWB do not exist to the best of our knowledge, although the influence of nature on the general SWB has been studied by researchers. The number of studies investigating the relationships between nature, especially green spaces, and SWB is increasing. In the following sections, we argue that forest-related SWB can be a promising policy indicator.

1.3. SWB-Correlated Factors in the Literature

In the reviewed two studies, the SWB-correlated factors in the literature were investigated. We organized factors associated with SWB into the following four categories in order to review previous studies: natural capital, built or manufactured capital, human capital, and social or cultural capital [19].

1.4. Natural Capital

Several researchers have emphasized the importance of natural areas as natural capital contributing to SWB by estimating their economic value. For example, Kopman and Rehdanz identified correlations between the ratios of natural environments and SWB in European countries and analyzed their monetary values as marginal willingness-to-pay (WTP) estimates [20]. Ambrey and Fleming reported relationships between ecosystem services (i.e., visual amenity and biodiversity) and SWB in Australia, and estimated WTP values for the improvement of these services [21,22]. An inverted U-shaped (concave) relationship between the distance to urban green spaces and life satisfaction was found for Berlin and this relationship was replicated in other German cities in a separate study [23,24]. In another study, Tsurumi and Managi identified relationships between SWB and the distance from the green spaces from residential areas, and calculated the marginal WTP for green spaces in metropolitan areas in Japan [25]. Tsurumi et al. further investigated various well-being measures such as the Cantril ladder, life satisfaction, subjective happiness, affect balance, and mental health, and their relationships with diverse green spaces in a metropolitan area in Japan and suggested possible positive effect of greenery investments on SWB [26]. A national-level survey in Japan found that respondents living in areas with more plantation forests or open water had relatively higher levels of well-being [27]. Based on these results, the authors recommended additional investments in plantation forest management and open water areas. Apart from land use types, Holms et al. captured the negative impacts of bark beetle epidemics on SWB in the western US [28].

Several studies have further explored the impact of interactions, not only physical closeness, with nature on SWB. Jang et al. reported that the frequency of forest visits is more relevant to SWB than the distance to urban forests in South Korea [29]. It was found in Italy and the UK that longer and more frequent visits to urban green spaces improved the perceived benefits and well-being of respondents [30]. Another study confirmed that the types of activities during visits to forest areas ((a) reading, talking, socializing; (b) walking/exercising, and (c) contemplating the setting) influenced well-being differently [31]. Bieling et al. noted that practices (e.g., hiking, walking) and perceived relationships (e.g., naturalness, tranquility, accessibility) in natural settings were also important factors for SWB compared to physical factors (e.g., mountains, forests, water bodies), based on open-ended interviews in Germany and Austria [32].

Notably, several studies have examined urban–rural differences based on SWB. In secondary analysis of existing data, lower levels of urban sprawl were associated with higher SWB as measured by personal financial issues for individuals living in urban areas in the USA [33]. Carrus et al. investigated how urban residents perceived peri-urban natural areas in large- to medium-scale cities in Italy and found that biodiversity had a positive relationship with perceived restorative properties and self-reported benefits from urban and peri-urban green spaces [31]. The aforementioned studies mainly focused on urban resident perspectives. Thus, we highlighted the well-being of residents in rural areas in the second (upper watershed) study reviewed below.

1.5. Built or Manufactured Capital

The correlations between SWB and built or manufactured capital (e.g., roads, postal offices) in Japan were examined, and there were no statistically significant correlations while the authors cautioned that several urban respondents completing the online survey may already have sufficient levels of built capital [27]. Tsurumi et al. included convenience indicators (e.g., number of neighboring retail stores) as control variables and found no statistically significant correlations with SWB [25].

1.6. Human Capital/Social or Cultural Capital

SWB is positively correlated with education level, which is often used as a measure of human capital [17]. Several studies included the levels of education as control variables and found statistically significant coefficients for the variables [25–27].

Social relationships are also correlated with SWB [12], which can be identified with social capital. Several studies analyzed above include social relationships, such as participation in volunteering activities [27] and the number of people who can be relied upon [26], with a statistically significant correlation and insignificant results, respectively.

1.7. Demographic Factors

Demographic factors were included as control variables in various studies [16,17,20–27]. Age, the age (squared), income, sex, marriage, and jobs were found to correlate with SWB.

1.8. Methodologies

All the aforementioned empirical studies involved survey or interview responses regarding SWB and used those responses as SWB indicators. However, MacKerron and Mourato utilized an innovative method by using a smartphone application to periodically collect daily life SWB data, as well as the GIS location when respondents reported their SWB, making more frequent measurement possible [34]. As an additional exception to the physical surveys and interviews, Wei et al. used facial expressions on selfies on SNS, a possibly more objective representation of SWB, to measure the satisfaction levels of visitors to forest parks in Chinese cities [35].

1.9. Structure of Paper

In summary, previous research empirically investigated the relationships between SWB and green spaces, such as forests, agricultural lands, and parks. They did not explicitly consider SWB or forest SWB as forest policy indicators. Based on two studies regarding forest SWB to be discussed in the below, we examined why and how forest SWB can be an appropriate policy indicator in the age of urbanization and forest restoration. Sections 2 and 3 review the summary results of two studies that examined forest SWB across the entire Yasu River watershed and at the upper Yasu River watershed in Japan. In Section 4, we argue that forest SWB can be utilized as a policy indicator, and finally, Section 5 provides a conclusion which also discusses possible future studies.

2. Study of Forest SWB across the Entire Yasu Watershed (Study 1)

In the current study, a questionnaire survey was distributed among Japanese residents living in a region with a medium-sized watershed [36]. A watershed represents a natural unit for forest management in the country, as indicated by regional forest management plans being based on watershed units. This watershed was selected because it has a densely populated lower watershed and a less populated upper watershed, which is typical in Japan [37]. Furthermore, the watershed residents' characteristics are representative of the Japanese population in terms of familiarity with forests, occasions upon which they visited forests, and their needs, based on comparisons of a local questionnaire with national surveys [38].

The study site, the Yasu River watershed, is located in Shiga Prefecture near Kyoto in central Japan (Figure 1A–C). Table 3 presents the descriptions of this watershed.

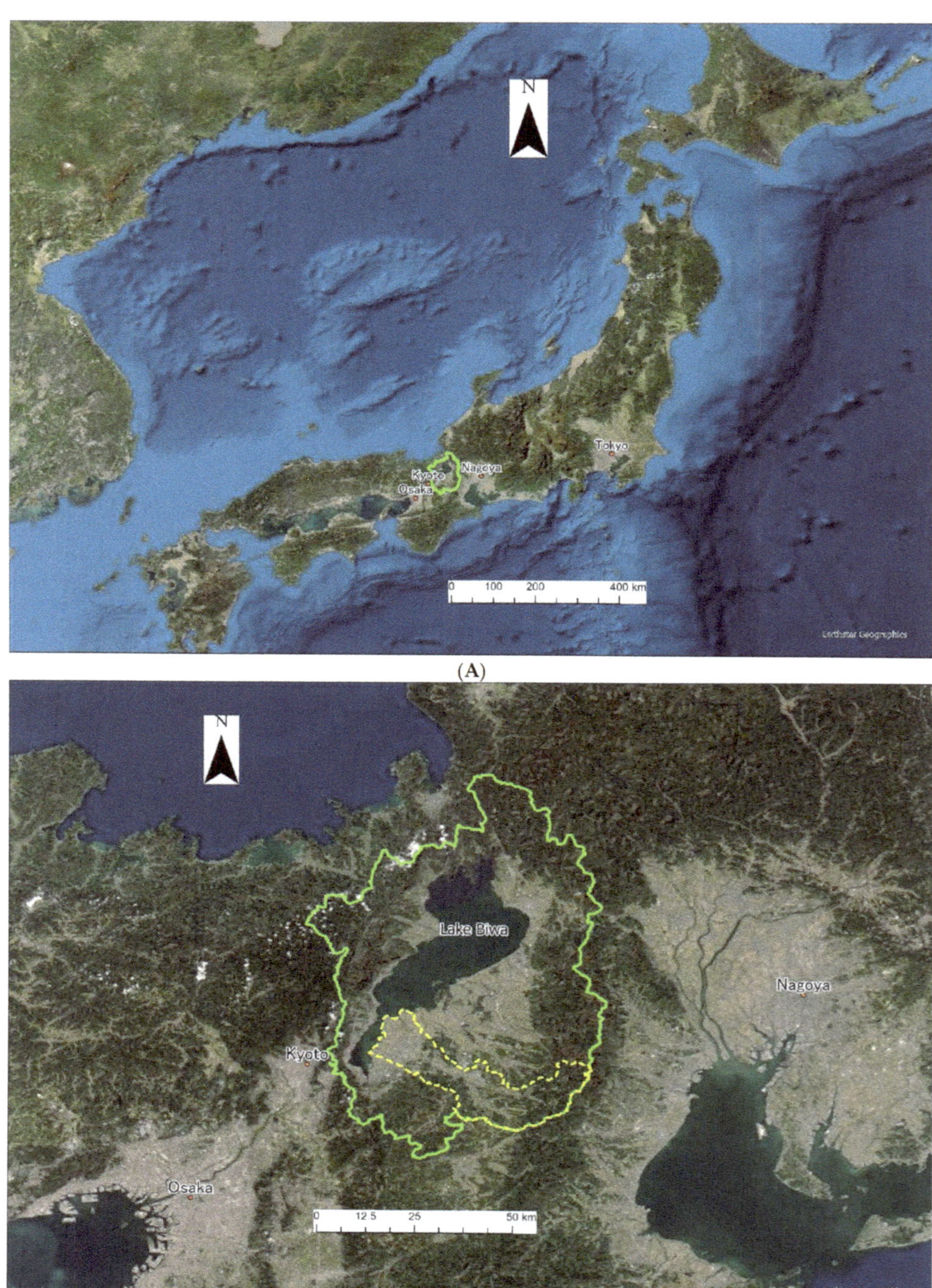

(B)

Figure 1. *Cont.*

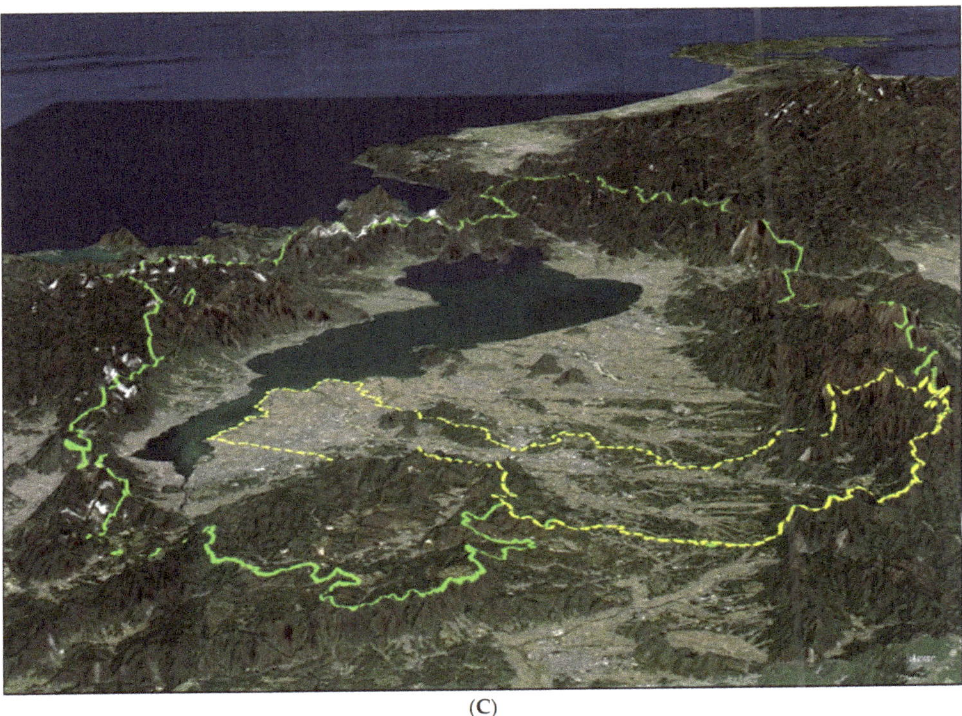

(C)

Figure 1. (**A**) Map of the study site (green line indicates Shiga Prefecture). All maps were drawn using ArcGIS Pro. (**B**) Map of the study site (green line indicates Shiga Prefecture, and yellow dotted line indicates the Yasu River watershed) The shapefile for the Yasu River watershed was provided by Dr. Satoshi Asano. (**C**) Map of the study site (3D representation).

Table 3. Yasu River watershed [36,39].

Topics	Descriptions
Coverage and size	The Yasu River tributary to Lake Biwa, the largest lake in Japan, is 65.3 km long with a watershed area of 387.0 km^2.
Society and economy	The watershed covers six municipalities with a combined population of 479,000 in 2015. The downstream areas consist of urban/rural mixed land with thriving commercial and industrial sections that capitalize on the advantages of the major railroads and motorways connecting the eastern and western areas of the country. The upstream areas are also urbanized to a lesser extent; most of this area is rural, with forestry (timber production) activities occurring within.
Forests	The total forested area in the six cities consists of 39,902 ha, comprising 51% of the total land area, while the forest ratios range from 6% to 55% among the six cities. 49% are plantation forests, and the ratios of plantation forests among the six cities vary from 0% to 67%. Forests owned by households are the largest category in terms of ownership (44%), followed by corporate (7%), and national forests (7%) in the six cities. The forests in the watershed are classified as temperate and are composed of (i) plantation forests dominated by Japanese cedar (*Cryptomeria japonica*) and cypress (*Chamaecyparis obtusa*), and (ii) natural forests dominated by konara oak (*Quercus serrata*).

Questionnaires were mailed in February and March 2016 to 34,691 households in 81 randomly selected postal codes within the study area. The questionnaire, written in Japanese, asked about SWB, relationships with others (social capital), nature (forest-related activities), and other aspects of everyday life. A total of 3,220 questionnaires were returned with a 9.3% response rate.

The average age of the respondents was 65, in comparison to residents' average age among the six cities ranging from 40 to 46 as of 2015. Respondents were 35% female, whereas female residents in the six cities range from 48% to 51% as of 2015 [40].

Forest SWB was measured using one item intended to assess affective evaluations towards local forests. The SWB specifically related to local forests was measured by the responses to the following statement: "I feel happy when I see local mountains." The response options ranged from 1 = "I completely disagree" to 5 = "I strongly agree." The term "mountains," which is interchangeable with "mountain forests" in Japanese, was used as vernacular to question the feelings regarding forests, which are located mainly in mountainous areas in Japan [41]. The distribution of forest SWB responses is presented in Figure 2 with the average score in brackets. Responses were widely distributed, indicating that there was sufficient individual variation.

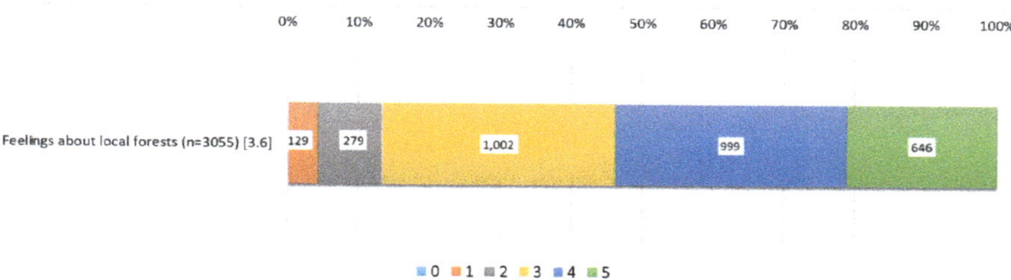

Figure 2. Forest subjective well-being (SWB) in the entire watershed study (feelings regarding local forests) [36]. The authors gave permission to use this chart.

Ordinary least squares (OLS) regressions were performed in STATA to identify significant associations ($p < 0.10$) between forest SWB, demographic factors, and (i) natural, (ii) man-made, and (iii) social capitals, with positive and negative coefficients indicating the direction of these associations (Table 4). In the written explanation of the results to follow, the numbers for each variable in Table 4 will be used in text to refer to their corresponding variables.

Table 4. Summary results of regression analyses of the entire watershed study: explanatory variables with positive and negative statistically significant coefficients ($p < 0.10$) * [36].

Dependent Variables	Explanatory Variables: Positive Coefficients	Explanatory Variables: Negative Coefficients
	Entire Watershed	
1 Feeling regarding local forests	1 Subjective health 2 Age (squared) 3 Social interaction 4 Working in forestry 5 Working in fishery 6 School education level 7 Leisure activities in forests 8 Seeing forests from homes 9 Engagement with wildlife	10 Working in business 11 No engagement with forests 12 Interaction terms between forest ratios and forest-related activities (leisure, observing from home, and engagement with wildlife)
	(no relation found) Areal ratio of forests in the area the respondent lives Areal ratios of natural or plantation forests in the area the respondent lives	

* Explanatory variables representing built or manufactured capital such as hospitals are omitted from this table for simplicity.

Considering the groups of variables, certain demographic variables have positive or negative correlations with the indicators.

- Subjective health (dependent variable 1–explanatory variable 1; 1–1) and age squared (1–2) were positively correlated with forest SWB.
- The professional category of respondents was also associated with forest SWB such that individuals working in forestry (1–4) and fishery (1–5) had higher forest SWB and those working in businesses (1–10) had lower forest SWB relationships.

Next, we considered variables relating to respondents' behavior regarding forests.

- Engagement with wildlife (1–9) and recreational activities (e.g., leisure) (1–7) had positive relationships with forest SWB.
- There was a significant interaction between the ratio of forest ratios and forest-related activities (1–12) predicting forest SWB. This indicates that respondents who lived in forested areas perceived lower forest SWB from forest-related activities than did those in less forested areas.

Unexpectedly, there was no significant associations between the physical presence of forests and forest SWB.

- The forest ratios of the respective postal areas where respondents resided did not correlate with forest SWB.

The adjusted R^2 value for the models with dependent variable 1 (forest SWB) was 0.156. The F-statistic p-value of the corresponding OLS was less than 0.0001. The adjusted R^2 values for models with a general SWB were larger than those for the forest SWB (0.360).

3. Study of Forest SWB within the Upper Yasu Watershed (Study 2)

A second study was conducted on the upper watershed, a subsection within the first study site [42]. The purpose of this second survey was to identify patterns of responses among populations that have more intense ties with forests. The upper watershed areas have higher ratios of forested areas, whose residents historically experience active forestry activities. The city in the upper watershed includes 81% of the all the six watershed cities [39]. Considering the entire watershed survey, it was found that residents in forested areas were less likely to derive forest SWB from forest-related activities, which is another motivation for this study.

Questionnaires were mailed from January through April 2018 to 6559 households in all postal codes of two upper watershed areas of the study area. Similar to the previous questionnaire, SWB, relationships (social capital), nature (forest-related activities), and other aspects of everyday life were questioned. A total of 1457 questionnaires were returned, with a response rate of 17.2%. The average age of the respondents was 59, consisting of 39% of female respondents while the average age of citizens of the upper watershed city was 46 and the ratio of female population was 50% [40].

In this study, forest SWB was measured using five items. Similar to study 1, the first item asked about personal affective evaluation towards local forests. The other four items assessed satisfaction, fulfillment (eudaimonia), positive affect, and negative affect, based on the Organization for Economic Co-operation and Development (OECD) guidelines for the measurement of SWB [43].

Forest satisfaction: Respondents were asked about their satisfaction using an 11-point scale, which ranged from "completely dissatisfied" (0) to "completely satisfied" (10) ("How satisfied are you with your current relationships with forests?"). Forest fulfillment (eudaimonia): Forest fulfillment was assessed on an 11-point scale ranging from "do not feel at all" (0) to "feel strongly" (10) ("How much worth, fulfilment, or sense of accomplishment do you feel regarding your relationships with forests?"). This question concerns eudaimonia, a concept first indicated by Aristotle in his Nicomachean Ethics, in which he asserts that people are happy not by feeling pleasure (hedonia), but by leading virtuous lives [44,45].

Positive and negative affect (feelings): Next, respondents were asked "How much have you experienced the following feelings during your experiences regarding forests?" For this item, respondents were provided the following list of feelings: forward-looking, backward-looking, pleasant, not pleasant, happy, sad, fearful, joyful, angry, satisfied, proud, shameful, awe, and respect. For each feeling, respondents were asked to choose one option on a five-point scale: 1 = "very rare," 2 = "rare," 3 = "sometimes," 4 = "frequently," and 5 = "very frequently." Based on the results of a factor analysis, by choosing items with factor loadings greater than 0.5, the authors determined the scores of the positive and negative affects by adding the scores of the five items for the positive affect (i.e., forward-looking, pleasant, happy, joyful, and satisfied), and the scores of two items for the negative affect (i.e., backward-looking and unpleasant).

Figure 3 presents the distributions of responses to the forest SWB items. The values in brackets indicate the average scores. Similar to the previous assessment, the responses are distributed widely, indicating sufficient variation in responses.

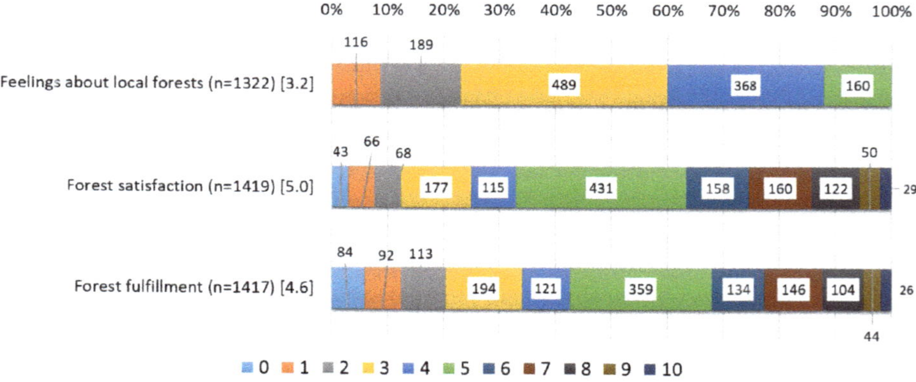

Figure 3. Forest SWB in the upper watershed study [42]. The authors gave permission to use this chart.

Figure 4 presents the distribution of forest SWB evaluated by respondents considering their affect (feelings). The values in brackets indicate the average. In general, respondents reported a higher frequency for positive affect than for negative affect. Nearly all indicators for the positive affect, such as "Forward-looking," "Pleasant," and "Happy," have higher scores than the mid-point (3.0) between 1 and 5; except for "Satisfied" [2.9] and "Proud" [2.6]. All the indicators for the negative affect, such as "Backward-looking," "Not Pleasant," "Sad," "Fearful," "Angry," and "Shameful," have lower scores than the mid-point (3.0). Although the number of responses in the second survey was 1457, the number of responses to complete the emotion items ranged from 647 to 842, because respondents who did not engage with forests were instructed not to answer these questions.

The validity of these measures was examined [42]. These measures are based on the manual for measuring SWB by the OECD [43]. While these measures concentrate on a specific domain, that is, relationships with forests, this treatment is justified by studies in psychology, in which a domain-specific SWB, such as the job and marriage of a participant, are measured [43]. Construct validity was verified using confirmatory factor analysis; the results obtained were satisfactory [42].

Ordinary least squares (OLS) regressions were performed in STATA to identify significant associations ($p < 0.10$) between forest SWB, demographic factors, and (i) natural, (ii) man-made, and (iii) social capitals, with positive and negative coefficients indicating the direction of these associations (Table 5). In the written explanation of the results to follow, the numbers for each variable in Table 5 will be used in text to refer to their corresponding variables.

Table 5. Summary results of regression analyses in the upper watershed study: explanatory variables with positive and negative statistically significant coefficients * ($p < 0.10$) [42].

Dependent Variables	Explanatory Variables: Positive Coefficients	Explanatory Variables: Negative Coefficients
	Upper watershed	
2 Feelings regarding local forests	1 Subjective health 2 Age (squared) 3 Female 4 Number of family members 5 School education level 6 Social interaction within local community 7 Fishing/collecting mountain vegetables	8 Age 9 Working in business 10 Part-time job 11 Student 12 Ownership of forests
3 Forest satisfaction	1 Subjective health 2 Working in agriculture 3 Climbing/skiing 4 Fishing/collecting mountain vegetables 5 Management of privately owned forest	6 Student 7 Ownership of forests
4 Forest fulfillment	1 Subjective health 2 Working as a civil servant 3 Working in agriculture 4 Working in forestry 5 Climbing/skiing 6 Management of privately owned forest 7 Management of forests as a volunteer	8 Female 9 Student 10 Ownership of forests
5 Positive affect	1 Subjective health 2 Female 3 Interaction with community members 4 Camping 5 Climbing/skiing 6 Observing animals and plants 7 Wood-working	8 Part-time job 9 Management of community forests 10 Ownership of forests
6 Negative affect **	1 Female 2 Climbing/skiing 3 Fishing/collecting mountain vegetables 4 Observing animals and plants 5 Plantation ratio of forests owned	6 Ownership of forests
All five dependent variables	(No relation found) Areal ratio of forests in the area the respondent lives	

* Variables indicating built or manufactured capital, such as hospitals, are omitted from this table for simplicity. ** As the dependent variable was reversed, "positive" indicates that an increase in the explanatory variable leads to a decrease in the negative affect, and vice versa.

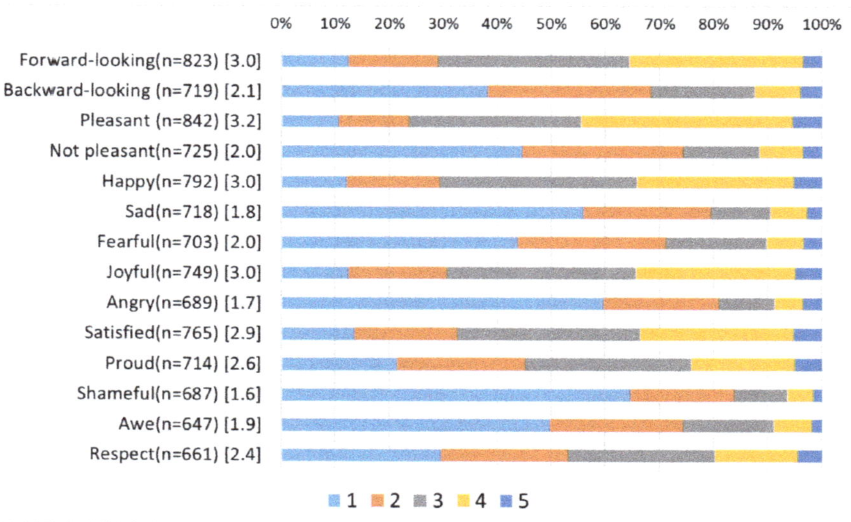

Figure 4. Forest SWB (positive and negative affect) in the upper watershed study [42]. The authors gave permission to use this chart.

Considering the groups of variables, certain demographic variables have positive or negative correlations with the indicators.

- Subjective health was positively correlated with forest SWB (2–1, 3–1, 4–1, 5–1).
- Female respondents were more likely to report higher forest SWB (2–3, 5–2, 6–1).
- Age correlated with forest SWB (2–2, 2–8). As the age terms and the age squared had negative and positive coefficients, respectively, in the "2 Feeling regarding local forests" analysis for the upper watershed survey, age and forest SWB had a curvilinear, U-shaped relationship with the lowest point at a positive age. The authors determined that the age of 53 had the lowest point.
- Respondent jobs had positive and negative relationships with forest SWB. Individuals working in forestry (4–4) and agriculture (3–2, 4–3) had positive relationships, while those working in business (2–9) and studying at school (2–11, 3–6, 4–9) had negative relationships.

Next, we consider the variables indicating respondent behavior regarding forests.

- Observing animals and plants (5–6, 6–4) had a positive relationship with forest SWB.
- Recreational activities, such as climbing/skiing (3–3, 4–5, 5–5, 6–2), and fishing/collecting mountain vegetables (2–7, 3–4, 6–3) were positively related to forest SWB.
- Management of privately owned forests (3–5, 4–6) had a positive relationship with forest SWB, and management of community forests (5–9) had a negative relationship.
- Ownership of forests (2–12, 3–7, 4–10, 5–10, 6–6) was negatively related to forest SWB.

Unexpectedly, the authors did not identify a relationship between the physical presence of the forests and forest SWB.

- The forest ratios of the respective postal areas where the respondents resided did not correlate with the forest SWB.

The adjusted R^2 values for models with dependent variables 2–6 (forest SWB) were 0.103, 0.116, 0.151, 0.160, and 0.107, respectively. All F-statistic p-values of the corresponding OLS were less than 0.0001. The adjusted R^2 values for models of general SWB were larger than those for the forest SWB (0.415).

4. Discussion

The reviewed studies suggest that forest SWB varied among respondents and correlated with respondents' engagement with local forests, such as forest-related activities and forest ownership. Forest ownership negatively correlated with forest SWB, indicating owner dissatisfaction with the unprofitability of timber producing forestry in contemporary Japan [42].

4.1. Rationales for Forest SWB as a Policy Indicator

In addition to these findings, based on the following four rationales, we argue that forest SWB can be an additional and promising policy indicator that complements the existing physically oriented indicators focusing on areas and volumes of forests.

The first rationale is that the direct measurement of well-being is preferable to indirect measurements, such as the gross domestic product (GDP). In the field of economics, income or GDP is used as an approximate measure for estimating individual well-being. Although researchers or policymakers frequently use economic indicators such as household income or GDP, such indicators neglect significant aspects of the forest ecosystem services. More specifically, individuals working in forestry constitute a relatively small portion of the population in Japan. In the entire watershed survey, only 0.3% of respondents worked in forestry and 1.4% worked in forestry in the upper watershed survey. Owing to the low profitability of forest management, few individuals work in forestry, even in the upper watershed. However, a significant portion of the respondents in the upper watershed (42%) owned forests. Furthermore, income from forestry constitutes only 0.04% of the GDP in Japan as of 2018 [46]. Only using economic indicators masks the important aspects of forest ecosystem services outside of income, such as venues for recreational activities or the embodiment of family traditions "that are central to quality of life and cultural identity" [47]. Engagement with animals and plants presents a positive correlation with forest SWB among some of the OLS analyses; mountain climbing and management of privately owned forests also had a positive relationship with forest SWB in the upper watershed study. An assessment relying on income measures overlooks these aspects of human–forest interactions because they represent only an insignificant amount in terms of money.

Second, SWB captures the quality of human interactions with forests, an aspect that is often missing in quantitative indicators, such as in the Montreal Process and Pan-European Process for Sustainable Forest Management. Here, we examine these indicators as international standards for sustainable forest management because the international community of policymakers and researchers of intergovernmental panels agreed on these, based on up-to-date knowledge of forest management and the practicality of their applications. For example, the Montreal process includes the following criteria:

Criterion 1: Conservation of biological diversity.

Criterion 2: Maintenance of the productive capacity of forest ecosystems.

Criterion 3: Maintenance of the forest ecosystem health and vitality.

Criterion 4: Conservation and maintenance of soil and water resources.

Criterion 5: Maintenance of forest contributions to global carbon cycles.

Criterion 6: Maintenance and enhancement of long-term multiple socioeconomic benefits.

Criterion 7: Legal, institutional, and economic frameworks for forest conservation and sustainable management.

Criterion 6 reflects the quality of citizen interactions with the forests. This aspect is becoming more significant in urbanized societies because urban residents do not have traditional relationships with forests, such as harvesting trees, mountain vegetables, and mushrooms. A national report from Japan on this process, the "State of Japan's Forests and Forest Management—3rd Country Report of Japan to the Montreal Process" presents the national survey results of changes in public expectations of forests (ranking) for indicator 6.5.b: the importance of forests to people. Currently, the national government cannot sufficiently monitor the quality of citizen interactions with forests because there is no verified method. Forest SWB may be an indicator of this aspect.

Third, SWB could identify inequalities between populations regarding access to or use of forest ecosystem services. As demonstrated by the reviewed studies, forest ownership, which is more prevalent in the upper watershed, unexpectedly presented a negative correlation with forest SWB. In contrast, urban residents are more likely to enjoy forest ecosystem services. The low levels of forest SWB for forest owners may reflect that forest management heavily burdens them, given the low profitability of timber-producing forestry; on the other hand, urban residents could receive forest ecosystem services without paying a price for them, other than taxes. Such an asymmetric pattern of forest SWB between stakeholders can indicate inequalities surrounding free access to or use of the forest ecosystem services, that is, the public goods nature of certain forest ecosystem services, as well as the cost burden of managing forests.

The fourth rationale is that SWB is a more holistic indicator that can capture subjective perspectives of respondents. The current studies' surveys included not only the economically rational evaluation by the respondents, but also responses of "feelings" towards forests. Forest restoration is a global trend and we face new challenges for improving forests qualitatively. For example, we may further need more forests with giant trees, which inspire visitors aesthetically or spiritually. Forest SWB measurements could better capture the perceived ecological qualities of forests, as well as the psychological and sociological forest–human interactions, such as access to and use of forests.

The current form of forest SWB has some limitations. A thorough representation of the forest ecosystems and the interactions between forests and humans may be limited. Forest SWB is not a rich, fully realistic description of forests or forest–human interactions; rather, it could be considered as a policy "indicator". For example, body temperature alone does not indicate a complete representation of health; however, it is used as an indicator of health because it is useful for monitoring health. This example demonstrates that the simplistic nature of forest SWB does not necessarily invalidate its use as a policy indicator.

4.2. Possibilities and Challenges for Forest SWB

Furthermore, forest SWB can contribute to enhancing urban–rural cooperation. Policy entrepreneurs could identify opportunities for improving the SWB of urban residents by identifying activities that positively correlate with their forest SWB. Rural communities could provide such opportunities to urban residents and potentially receive some rewards (e.g., human resource and monetary resource) from them. Conversely, higher levels of forest SWB in certain rural communities may suggest novel ways to improve forest SWB with innovative lifestyles or activities. Researchers could identify urban–rural combinations to enhance mutually beneficial relationships among forest SWB. For example, policy measures, such as payment for ecosystem services (PES) and product certification would coordinate the relationship between forest owners and urban residents.

Forest SWB may have future applications for developing countries as well. Researchers and practitioners in developing countries may claim that forest SWB could provide useful information only for a certain group of industrialized countries where forest restoration has been achieved, and not for developing countries, where forest destruction and poverty are more urgent issues. However, as Mather suggests, one reason for restoring forests is the change in feelings of individuals [2], thus forest SWB should be monitored, and factors influencing forest SWB should be investigated also in developing countries.

In the field of conservation, overall SWB (not forest SWB) is considered to be a potential policy indicator in developing countries. Biedenweg and Gross-Camp proposed to incorporate well-being into conservation dialogues for the two following reasons: (1) conservation without considering the well-being of affected individuals will fail, and (2) environmental justice requires considering well-being, including the distribution of costs and benefits of conservation [48]. Social impact assessment with subjective well-being measurement in the Global South was also proposed [49]. Several empirical studies have assessed the relationship between forest management and SWB in developing countries. A study measured the subjective well-being of residents in an area of the Brazilian Amazon and found no correlation with the participation in logging projects [50]. Another study evaluated the impact of REDD+ on the subjective well-being of 4000 households in 130 villages in Brazil, Peru, Cameroon, Tanzania, Indonesia, and Vietnam [51]. Researchers and practitioners in the field of resource management may be able to gain a better understanding by learning from these examples and focusing on local contexts as well as scientific rigor with practical considerations.

Considering governmental policies in developing countries, the Bhutanese government monitors the well-being of individuals, which involves subjective aspects and relationships with the natural environment and sets a gross national happiness (GNH) level as the national goal [52,53]. These examples are worth examining for incorporating SWB into forest policies.

Here, we propose several possibilities indicating why the models presenting the forest SWB had lower R^2 values than those for the general SWB. This can help improve the analytical capabilities of models used for future studies. First, the models for forest SWB may have missing factors that determine the levels of forest SWB. For example, personality traits, such as extraversion or introversion, may exert a stronger influence on forest SWB than on general SWB. Second, the measurement of explanatory variables may be inappropriate. The intensity of forest-related activities was found to correlate with the general SWB or the evaluation of forest spaces for recreational purposes [29,31]. The current models assess the levels of forest-related activities as yes-or-no experiences during the past year. A change from binary to graded measurements may improve the explanatory power of the models. Third, forest SWB may involve more measurement errors than the general SWB. Respondents may have greater difficulty assessing their SWB in this specific domain of their lives. Future studies can explore these possibilities by including new explanatory variables and improving the measurement methods.

5. Conclusions

This study proposes forest SWB as a promising policy indicator based on measurement trials in Japan. Studies 1 and 2 suggest that there was variation in respondents' level of self-reported measures of forest SWB and forest SWB was significantly associated with some demographic variables and behaviors related to forest interactions. We further discussed the following rationales for using forest SWB as a policy indicator: (1) direct measurement of well-being—SWB is preferable to indirect measurements such as income levels; (2) well-being measures are influenced by respondents' interactions with forests; (3) inequalities among stakeholders can be identified; (4) SWB represents a holistic measurement.

Despite these novel findings and suggestions, challenges remain in establishing forest SWB as a policy indicator. Survey instruments for measuring forest SWB have not been standardized. Based on several measurement trials, we should ensure that survey instruments, especially survey questions, are valid and reliable. Efforts to improve the explanatory power of models explaining forest SWB should be increased. The mechanism for determining forest SWB should be further investigated. Although we attempted to control endogeneity by estimating the average treatment effect based on an endogenous treatment-regression model [36], other methods such as randomized controlled tests or the regression discontinuity design may provide more unbiased and efficient results. Measurements among different populations (e.g., different countries and regions) may reveal the generality or specificity of forest SWB and could contribute to revealing hidden mechanisms.

Author Contributions: All authors made contributions to the conception of this paper. T.T. and Y.U. designed the questionnaire instrument. T.T. led the questionnaire survey, data analysis and drafting, and Y.U., H.I. and N.O. critically reviewed drafts for intellectual content and revised them. All authors have read and agreed to the published version of the manuscript.

Funding: This research was funded by Research Institute for Humanity and Nature for the project, "Biodiversity-driven Nutrient Cycling and Human Well-being in Social-Ecological Systems (D06-14200119)" and JSPS KAKENHI Grant-in-Aid for Scientific Research (B) 15H02871 and (B) 20H03090.

Institutional Review Board Statement: This statement is not applicable to the current review article.

Informed Consent Statement: This statement is not applicable to the current review article.

Data Availability Statement: Data available in a publicly accessible repository.

Acknowledgments: Kimberly Bowen provided valuable comments and suggestions to the manuscript. We would like to thank Editage for editing and reviewing this manuscript for English language.

Conflicts of Interest: The authors declare no conflict of interest.

References

1. United Nations, Department of Economic and Social Affairs, P.D. *World Urbanization Prospects: The 2018 Revision (ST/ESA/SER.A/420)*; United Nations: New York, NY, USA, 2019; pp. 1–103.
2. Mather, A.S. The forest transition. *Area* **1992**, *24*, 367–379.
3. Mather, A.; Needle, C.L. The forest transition: A theoretical basis. *Area* **1998**, *30*, 117–124. [CrossRef]
4. MacDonald, H.; McKenney, D. Envisioning a global forest transition: Status, role, and implications. *Land Use Policy* **2020**, *99*, 104808. [CrossRef]
5. Ota, T. *Forest Saturation: Thinking about Changing National Land*; NHK Publishing: Tokyo, Japan, 2012; pp. 1–254.
6. Weible, C.M.; Sabatier, P.A. (Eds.) *Theories of the Policy Process*; Westview Press: New York, NY, USA, 2018; pp. 1–416.
7. IPBES. *Summary for Policymakers of the Global Assessment Report on Biodiversity and Ecosystem Services of the Intergovernmental Science-Policy Platform on Biodiversity and Ecosystem Services*; Díaz, S., Settele, J., Brondízio, E.S., Ngo, H.T., Guèze, M., Agard, J., Arneth, A., Balvanera, P., Brauman, K.A., Butchart, S.H.M., et al., Eds.; IPBES: Bonn, Germany, 2019; pp. 1–45.
8. Ministry of Agriculture, Forestry and Fisheries. *National Forest Plan (Approved by the Cabinet on October 16, 2018)*; Ministry of Agriculture, Forestry and Fisheries: Tokyo, Japan, 2018; pp. 1–30.
9. Akiyama, T. Attitudes of members of forest owner's associations toward forestry and challenges for the associations: Based on the result of questionnaire surveys. *Financ. Agric. For.* **2008**, *61*, 27–41.
10. Forestry Agency. *White Paper of Forests and Forestry (Heisei 29 (2017))*; Forestry Agency: Tokyo, Japan, 2018; pp. 1–246.
11. FAO. *Global Forest Resources Assessment 2015 (FRA 2015)*; FAO: Rome, Italy, 2015; pp. 1–244.
12. Eid, M.; Larsen, R.J. (Eds.) *The Science of Subjective Well-Being*; The Guilford Press: New York, NY, USA, 2008; pp. 1–546.
13. Diener, E. Subjective well-being: The science of happiness and a proposal for a national index. *Am. Psychol.* **2000**, *55*, 34–43. [CrossRef]
14. Diener, E.; Lucas, R.E.; Oishi, S. Subjective well-being: The science of happiness and life satisfaction. In *Handbook of Positive Psychology*; Snyder, C.R., Lopez, S.J., Eds.; Oxford University Press: Oxford, UK, 2002; pp. 63–73.
15. Kahneman, D.; Wakker, P.; Sarin, R. Back to Bentham? Explorations of experienced utility. *Quart. J. Econ.* **1997**, *112*, 375–405. [CrossRef]
16. Frey, B.S.; Stutzer, A. What can economists learn from happiness research? *J. Econ. Lit.* **2002**, *40*, 402–435. [CrossRef]
17. Dolan, P.; Peasgood, T.; White, M. Do we really know what makes us happy? A review of the economic literature on the factors associated with subjective well-being. *J. Econ. Psychol.* **2008**, *29*, 94–122. [CrossRef]
18. Millennium Ecosystem Assessment. Millennium Ecosystem Assessment. Available online: https://www.millenniumassessment.org/en/Global.html (accessed on 2 July 2020).
19. Costanza, R.; de Groot, R.; Braat, L.; Kubiszewski, I.; Fioramonti, L.; Sutton, P.; Farber, S.; Grasso, M. Twenty years of ecosystem services: How far have we come and how far do we still need to go? *Ecosyst. Serv.* **2017**, *28*, 1–16. [CrossRef]
20. Kopmann, A.; Rehdanz, K. A human well-being approach for assessing the value of natural land areas. *Ecol. Econ.* **2013**, *93*, 20–33. [CrossRef]
21. Ambrey, C.L.; Fleming, C.M. Valuing Ecosystem Diversity in South East Queensland: A Life Satisfaction Approach. *Soc. Indic. Res.* **2014**, *115*, 45–65. [CrossRef]
22. Ambrey, C.L.; Fleming, C.M. Valuing scenic amenity using life satisfaction data. *Ecol. Econ.* **2011**, *72*, 106–115. [CrossRef]
23. Bertram, C.; Rehdanz, K. The role of urban green space for human well-being. *Ecol. Econ.* **2015**, *120*, 139–152. [CrossRef]
24. Krekel, C.; Kolbe, J.; Wüstemann, H. The greener, the happier? The effect of urban land use on residential well-being. *Ecol. Econ.* **2016**, *121*, 117–127. [CrossRef]
25. Tsurumi, T.; Managi, S. Environmental value of green spaces in Japan: An application of the life satisfaction approach. *Ecol. Econ.* **2015**, *120*, 1–12. [CrossRef]
26. Tsurumi, T.; Imauji, A.; Managi, S. Greenery and Subjective Well-being: Assessing the Monetary Value of Greenery by Type. *Ecol. Econ.* **2018**, *148*, 152–169. [CrossRef]
27. Kunugi, Y.; Arimura, T.H.; Nakashizuka, T.; Oguro, M. Subjective well-being and natural capital in Japan—An empirical analysis using micro data. *Environ. Sci.* **2017**, *30*, 96–106.

28. Holmes, T.; Koch, F. Bark Beetle epidemics, life satisfaction, and economic well-being. *Forests* **2019**, *10*, 696. [CrossRef]
29. Jang, Y.-S.; Yoo, R.-H.; Lee, J.-H. The Effects of Visit Characteristics in Neighborhood Forest on Individual Life Satisfaction. *J. People Plants Environ.* **2019**, *22*, 677–690. [CrossRef]
30. Lafortezza, R.; Giuseppe Carrus, G.; Sanesi, G.; Clive Davies, C. Benefits and well-being perceived by people visiting green spaces in periods of heat stress. *Urban For. Urban Green.* **2009**, *8*, 97–108. [CrossRef]
31. Carrus, G.; Scopelliti, M.; Lafortezza, R.; Colangelo, G.; Ferrini, F.; Salbitano, F.; Agrimi, M.; Portoghesi, L.; Semenzato, P.; Sanesi, G. Go greener, feel better? The positive effects of biodiversity on the well-being of individuals visiting urban and peri-urban green areas. *Landsc. Urban Plan.* **2015**, *134*, 221–228. [CrossRef]
32. Bieling, C.; Plieninger, T.; Pirker, H.; Vogl, C.R. Linkages between landscapes and human well-being: An empirical exploration with short interviews. *Ecol. Econ.* **2014**, *105*, 19–30. [CrossRef]
33. Lee, W.H.; Ambrey, C.; Pojani, D. How do sprawl and inequality affect well-being in American cities? *Cities* **2018**, *79*, 70–77. [CrossRef]
34. MacKerron, G.; Mourato, S. Happiness is greater in natural environments. *Glob. Environ. Chang.* **2013**, *23*, 992–1000. [CrossRef]
35. Wei, H.; Hauer, R.J.; Chen, X.; He, X. Facial expressions of visitors in forests along the urbanization gradient: What can we learn from selfies on social networking services? *Forests* **2019**, *10*, 1049. [CrossRef]
36. Takahashi, T.; Asano, S.; Uchida, K.; Takemura, K.; Fukushima, S.; Matsushita, K.; Okuda, N. Effects of forests and forest-related activities on the subjective well-being of residents in a Japanese watershed: An econometric analysis through the capability approach. *For. Policy Econ.* **2021**. under review.
37. Ministry of Land, Infrastructure, Transport and Tourism. I.Land and Climate of Japan. Available online: https://www.mlit.go.jp/river/basic_info/english/land.html (accessed on 2 November 2020).
38. Takahashi, T. Attitudes of Residents in the Southern Shiga Watershed toward Forests: Investigation of its Structure based on a Questionnaire Survey. *J. Rural Probl.* **2009**, *45*. [CrossRef]
39. Shiga Prefecture. *Statistical Book of Shiga Prefecture. 2018. (Siga ken toukeisyo, Heisei 30 nendo, 2018 nendo)*; Shiga Prefecture: Otsu, Japan, 2020.
40. Ministry of Internal Affaiirs and Communication. *National Census 2015*; Ministry of Internal Affaiirs and Communication: Tokyo, Japan, 2017.
41. Naumann, N. *Yama no Kami (Yama no Kami: Die japanische Berggottheit)*; Gensosha: Tokyo, Japan, 1994.
42. Takahashi, T.; Uchida, Y.; Ishibashi, H.; Okuda, N. Factors affecting forest-related subjective well-being: A case study in the upper Yasu River watershed. *J. Jpn. For. Soc.* **2021**, *103*. in press.
43. OECD. *OECD Guidelines on Measuring Subjective Well-being*; OECD Publishing: Paris, France, 2013; pp. 1–265.
44. Ryff, C.D.; Keyes, C.L.M. The structure of psychological well-being revisited. *J. Pers. Soc. Psychol.* **1995**, *69*, 719. [CrossRef] [PubMed]
45. Aristotle. *Aristotle's Nicomachean Ethics*; The University of Chicago Press: Chicago, IL, USA, 2011.
46. Forestry Agency. *White Paper of Forests and Forestry (Reiwa 1 (2019))*; Forestry Agency: Tokyo, Japan, 2020.
47. IPBES. *Report of the Plenary of the Intergovernmental Science-Policy Platform on Biodiversity and Ecosystem Services on the Work of Its Seventh Session*; IPBES: Bonn, Germany, 2019.
48. Biedenweg, K.; Gross-Camp, N.D. A brave new world: Integrating well-being and conservation. *Ecol. Soc.* **2018**, *23*, 1–4. [CrossRef]
49. Woodhouse, E.; Homewood, K.M.; Beauchamp, E.; Clements, T.; McCabe, J.T.; Wilkie, D.; Milner-Gulland, E.J. Guiding principles for evaluating the impacts of conservation interventions on human well-being. *Philos. Trans. R. Soc. B Biol. Sci.* **2015**, *370*. [CrossRef]
50. Cooper, N.A.; Kainer, K.A. To log or not to log: Local perceptions of timber management and its implications for well-being within a sustainable-use protected area. *Ecol. Soc.* **2018**, *23*. [CrossRef]
51. Duchelle, A.E.; de Sassi, C.; Jagger, P.; Cromberg, M.; Larson, A.M.; Sunderlin, W.D.; Atmadja, S.S.; Resosudarmo, I.A.P.; Pratama, C.D. Balancing carrots and sticks in REDD+: Implications for social safeguards. *Ecol. Soc.* **2017**, *22*. [CrossRef]
52. Centre for Bhutan Studies. *A Compass Towards a Just and Harmonious Society 2015 GNH Survey Report*; Centre for Bhutan Studies: Timphu, Bhutan, 2016; pp. 1–334.
53. Centre for Bhutan Studies; GNH. *Happiness: Tranforming the Development Landscape*; Centre for Bhutan Studies: Timphu, Bhutan, 2017; pp. 1–475.

MDPI
St. Alban-Anlage 66
4052 Basel
Switzerland
Tel. +41 61 683 77 34
Fax +41 61 302 89 18
www.mdpi.com

Sustainability Editorial Office
E-mail: sustainability@mdpi.com
www.mdpi.com/journal/sustainability

www.ingramcontent.com/pod-product-compliance
Lightning Source LLC
LaVergne TN
LVHW070621170726
843515LV00004B/148

9783036518107